Astrophysics of Gaseous Nebulae

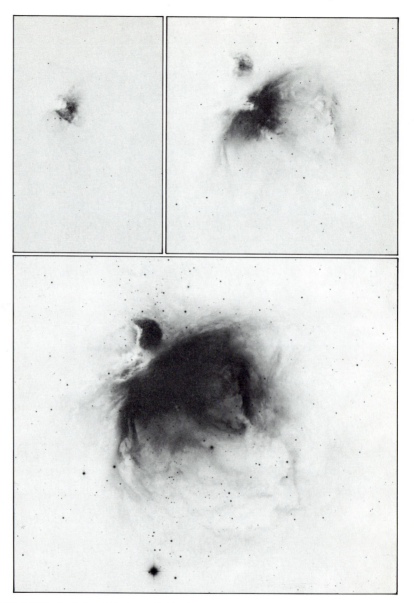

NGC 1976, the Orion Nebula, probably the most-studied H II region in the sky. Three different exposures, taken with the 48-inch Schmidt telescope, red filter and 103a-E plates, emphasizing Hα λ6563, [N II] λλ6548, 6583. The 5-second exposure (*upper left*) shows only the brightest parts of nebula; θ^1 Ori (the Trapezium), the multiple-exciting star, cannot be seen on even this short an exposure. In the 40-second exposure (*upper right*), the central parts are overexposed. The nebula to the upper left (northeast) is NGC 1982. Though it is partly cut off from NGC 1976 by foreground extinction, radio-frequency measurements show that there is a real density minimum between the two nebulae. The 6-minute exposure (*bottom*) shows outer fainter parts of nebula. Still fainter regions can be recorded on even longer exposures. Compare these photographs with Figure 5.7, which was taken in continuum radiation. (*Hale Observatories photographs.*)

Astrophysics of Gaseous Nebulae

Donald E. Osterbrock
Lick Observatory, University of California, Santa Cruz

W. H. Freeman and Company
San Francisco

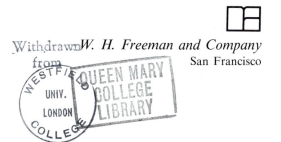

A Series of Books in Astronomy and Astrophysics

EDITORS: Geoffrey Burbidge
Margaret Burbidge

Cover: Planetary nebula in Aquarius.
Photograph by Hale Observatories.

Library of Congress Cataloging in Publication Data

Osterbrock, Donald E
    Astrophysics of gaseous nebulae.

    1.   Nebulae.   I.   Title.
QB855.087   523.1′135   74-11264
ISBN 0-7167-0348-3

Printed in the United States of America

1   2   3   4   5   6   7   8   9

Dedication

*To my friends at the University of Wisconsin
—fellow faculty members, postdocs, students,
co-workers, and especially the PhDs who did
their theses with me—during the fifteen years
I learned so much of what went into this
book from working with them.*

Contents

Preface

Probably no subject in astrophysics has grown so rapidly in the past twenty-five years, nor contributed so much to our understanding of the universe, as the study of gaseous nebulae. The investigation of the physical processes in gaseous nebulae by Menzel, Goldberg, Aller, Baker, and others before World War II led naturally to the theory of H II regions developed by Strömgren, which, in its turn, stimulated the observational work on the spiral arms of M 31 and other galaxies by Baade, and on the spiral arms of our Galaxy by Morgan. Much of what we know of the galactic structure and dynamics of Population I objects, of the abundances of the light elements, and of the ultraviolet radiation emitted by hot stars has been learned from the study of H II regions, while observations of planetary nebulae have led to considerable knowledge of the abundances in highly evolved old objects, the galactic structure and dynamics of Population II objects, and the final stages of evolution of stars with masses of the same order as the sun's mass. Though until a few years ago, only optical radiation was observed from nebulae, more recently radio-frequency and infrared measurements have added greatly to our understanding. No doubt ultraviolet measurements from artificial satellites will also soon become important.

Stimulated by these observational developments, many new theoretical advances have been made in the last twenty-five years. Because gaseous nebulae have such low densities, and consequently such small optical depths, the physical processes that occur in nebulae are often almost directly observed in the emergent radiation from them. Thus the theory of gaseous nebulae is not overcomplicated, and the new advances have come largely from advances in our knowledge of atomic processes—newly calculated collisional-excitation cross sections, recombination coefficients, and so on. These atomic parameters in turn are available today because of the availability of large digital computers, with which Schrödinger's equation can be numerically integrated in approximations developed largely by physicists working toward the goal of interpreting nebulae; much of this work has been done by Seaton and his collaborators.

Though the field is growing and interesting, and many new workers are coming into it from other branches of physics and astronomy, the available technical books on gaseous nebulae are, to my mind, somewhat out of date. I felt that there was a definite need for a new monograph on gaseous nebulae, and was encouraged in this feeling by other astronomers. The purpose of this book, therefore, is to provide a text that can be used at the introductory graduate level, or that can be used by research workers who want to familiarize themselves with the ideas, results, and unsolved problems of gaseous nebulae. This book is based on an ever-changing course that I have given several times to first- and second-year astronomy and physics graduate students at the University of Wisconsin, and that I gave to astronomy students at Yerkes Observatory once. It represents the material I consider necessary to understand papers now being published, and to do research on gaseous nebulae at the present time.

Naturally, it is impossible to include all the research that has been done on nebulae, but I have tried to select what I consider the basic theory, and to include enough observational and theoretical results to illustrate it. Many tables are included so that the book can be used to calculate actual numerical results; these tables are often selections of the most essential parts of more complete tables in the original references.

The reader of this book is assumed to have a reasonably good preparation in physics and some knowledge of astronomy and astrophysics. The simplest concepts of radiative transfer are used without comment, since almost invariably the reader has encountered this material before studying gaseous nebulae. The collision cross sections, transition probabilities, energy levels, and other physical results are taken as known quantities; no attempt is made in the book to sketch their derivation. When I teach this material myself I usually include some of these derivations, but they must be carefully linked to the quantum mechanics text in use at the institution at the time the course is taught. By omitting the radiative-transfer and quantum-mechanics material, I have been able to concentrate a good deal of material on nebulae into the book.

I have not inserted references throughout the text, partly because I feel they break up the flow and therefore interfere with understanding, and partly because in many cases the book contains the distillation of a mixture formed from a number of papers; moreover, there is no obvious place where many of the references should go. Instead, I have collected the references for each chapter in a separate section at the end, together with explanatory text. I urge the reader to look at these references after he has completed reading each chapter, and to study further those papers that deal with subjects in which he is particularly interested. Practically all the references are to the American and English literature with which I am most familiar; this is also the material that will, I believe, be most accessible to the average reader of this book.

I should like to thank many people for their aid in the preparation of this book. I am particularly grateful to my teachers at the University of Chicago, S. Chandrasekhar, W. W. Morgan, T. L. Page, and B. Strömgren, who introduced me to the beauties and intricacies of gaseous nebulae. I learned much from them, and over the years they have continued to encourage me in the study of these fascinating objects. I am also very grateful to R. Minkowski, my colleague and mentor at the Mount Wilson and Palomar Observatories, as it was then, who encouraged me and helped me to go on with observational work on nebulae.

I am also extremely grateful to my colleagues who read and commented on drafts of various chapters of the book—D. P. Cox, G. A. MacAlpine, W. G. Mathews, J. S. Miller, and M. Peimbert—and above all, my ex-student and friend for twenty years, J. S. Mathis, who read and corrected the entire first seven chapters. Though they found many errors, corrected many misstatements, and cleared up many obscurities, the ultimate responsibility for the book is mine, and the remaining errors that will surely be found after publication are my own. I can only repeat the words of a great physicist, "Listen to what I mean, not to what I say"; if the reader finds an error, it will mean he really understands the material, and I shall be glad to receive a correction.

I am particularly indebted to Mrs. Carol Betts, who typed and retyped many heavily corrected drafts of every chapter in the book, and to Mrs. Helen Hay, who typed the final manuscript and prepared the entire book for the publisher, for the skill, accuracy, care, and, above all, dedication with which they worked on this book. I am also most grateful to Mrs. Beatrice Ersland, who organized the Washburn Observatory office work over a three-year period so that the successive drafts could be typed, and to my wife, who carefully proofread the entire manuscript.

My research on gaseous nebulae has been supported over the years by the Research Corporation, the Wisconsin Alumni Research Foundation funds administered by the University of Wisconsin Graduate School, the John Simon Guggenheim Memorial Foundation, the Institute for Advanced Studies, and above all by the National Science Foundation. I am grateful to them all for their generous support. Much of my own research has gone into this book; without doing that research I could never have written the book.

I am especially grateful to E. R. Capriotti, R. F. Garrison, W. W. Morgan, G. Münch, and R. E. Williams, who supplied original photographs for this book. Many of the other figures are derived from published research papers; I am very grateful to the officers of the International Astronomical Union, of the Royal Astronomical Observatory, and of the European Southern Observatory for permission to use their figures, and also to the Hale Observatories, Lick Observatory, the University of Chicago Press, the *Reviews of Modern Physics,* and the *Publications of the Astronomical Society of the Pacific,* as well as all the individual authors whose papers appeared in these journals.

Madison, Wisconsin   *Donald E. Osterbrock*
August 1973

Astrophysics of Gaseous Nebulae

1

General Introduction

1.1   Gaseous Nebulae

Gaseous nebulae are observed as bright extended objects in the sky. Those with the highest surface brightness, such as the Orion Nebula (NGC 1976) or the Ring Nebula (NGC 6720), are easily observed on direct photographs, or even at the eyepiece of a telescope. Many other nebulae that are intrinsically less luminous or that are more strongly affected by interstellar extinction are faint on ordinary photographs, but can be photographed on long exposures with filters that isolate a narrow wavelength region around a prominent nebular emission line, so that the background and foreground stellar and sky radiations are suppressed. The largest gaseous nebula in the sky is the Gum Nebula, which has an angular diameter of the order of 30°, while many familiar nebulae have sizes of the order of one degree, ranging down to the smallest objects at the limit of resolution of the largest telescopes. The surface brightness of a nebula is independent of its distance, but more distant nebulae have (on the average) smaller angular size and greater interstellar extinction, so the nearer members of any particular type of nebula tend to be the most-studied objects.

Gaseous nebulae have an emission-line spectrum. This spectrum is dominated by forbidden lines of ions of common elements, such as [O III] λλ4959, 5007, the famous green nebular lines once thought to indicate the presence of the hypothetical element nebulium; [N II] λλ6548, 6583 in the red; and [O II] λλ3726, 3729, the ultraviolet doublet, which appears as a blended λ3727 line on low-dispersion spectrograms of almost every nebula. In addition, the permitted lines of hydrogen, Hα λ6563 in the red, Hβ λ4861 in the blue, Hγ λ4340 in the violet, and so on, are characteristic features of every nebular spectrum, as is He I λ5876, which is considerably weaker, while He II λ4686 occurs only in higher-excitation nebulae. Long-exposure spectrograms, or photoelectric spectrophotometric observations extending to faint intensities, show progressively weaker forbidden lines, as well as faint permitted lines of common elements, such as C II, C III, C IV, O II, and so on. The emission-line spectrum, of course, extends into the infrared and presumably also into the ultraviolet, but as of this writing no nebular spectrogram extending below 3000 A has been obtained, except for quasi-stellar objects with very large red shifts.

Gaseous nebulae also have a weak continuous spectrum, consisting of atomic and reflection components. The atomic continuum is emitted chiefly by free-bound transitions, mainly in the Paschen continuum of H I at λ > 3648 A, and the Balmer continuum at λ < 3648 A. In addition, many nebulae have reflection continua consisting of starlight scattered by dust. The amount of dust varies from nebula to nebula, and the strength of this continuum fluctuates correspondingly. In the infrared, the nebular continuum is largely thermal radiation emitted by the dust.

In the radio-frequency region, emission nebulae have a reasonably strong continuous spectrum, mostly due to free-free emission or bremsstrahlung of thermal electrons accelerated in Coulomb collisions with protons. Superimposed on this continuum are weak emission lines of H, such as 109α at 6 cm, resulting from bound-bound transitions between very high levels of H. Weaker radio recombination lines of He and still weaker lines of other elements can also be observed in the radio region, slightly shifted from the H lines by the isotope effect.

1.2   Observational Material

Practically every observational tool of astronomy can be and has been applied to the study of gaseous nebulae. Because nebulae are low-surface-brightness, extended objects, the most effective instruments for studying them are fast, wide-field optical systems. For instance, large Schmidt cameras are ideal for direct photography of gaseous nebulae, and many of the most familiar pictures of nebulae, including several of the illustrations in this book, were taken with the 48-inch Schmidt telescope at Palomar Observatory. The finest small-scale detail in bright nebulae is best shown on photographs taken with longer focal-length instruments, as other illustrations taken with the 200-inch F/3.67 Hale telescope and with the 120-inch F/5 Lick telescope show.

Though the brighter nebulae are known from the early visual observations, many fainter-emission nebulae have been discovered more recently by systematic programs of direct photography, comparing an exposure taken in a narrow wavelength region around prominent nebular lines (most often Hα λ6563 + [N II] $\lambda\lambda$6548, 6583) with an exposure taken in another wavelength region that suppresses the nebular emission (for instance, $\lambda\lambda$5100–5500). Other small nebulae have been found on objective-prism surveys as objects with bright Hα or [O III] emission lines, but faint continuous spectra.

Much of the physical analysis of nebulae depends on spectrophotometric measurements of emission-line intensities, carried out either photographically or photoelectrically. The photoelectric scanners have higher intrinsic accuracy for measuring a single line, because of the higher quantum efficiency of the photoelectric effect and the linear response of a photoelectric system, while the photographic plate has the advantage of "multiplexing," or recording very many picture elements (many spectral lines in this case) simultaneously. However, many-channel photoelectric systems, or image-dissector systems, which retain the advantages of photoelectric systems while adding the multiplexing property formerly available only photographically, are now being developed and put into use. A fast nebular spectrophotometer can be matched to any telescope, but the larger the aperture of the telescope, the smaller the size of the nebular features that can be accurately measured, or the fainter the small nebulae (those with angular size smaller than the entrance slit or diaphragm) that can be accurately measured.

Radial velocities in nebulae are measured on slit spectrograms. Here again, a fast spectrograph is essential to reach low-surface-brightness objects, and a large telescope is required to observe nebular features with small angular size.

Nebular infrared continuum measurements can be made with broad-band photometers and radiometers, using filters to isolate various spectral regions. Spectrophotometry of individual infrared lines requires better wavelength resolution than can be achieved with ordinary filters, and Fabry-Perot interferometers and Fourier spectrographs have just begun to be used for these measurements. In the radio region, filters are used for high wavelength resolution. In all spectral regions, large telescopes are required to measure features with small angular size; in particular, in the radio region long-baseline interferometers are required.

1.3   Physical Ideas

The source of energy that enables emission nebulae to radiate is, in almost all cases, ultraviolet radiation from stars involved in the nebula. There are one or more hot stars, with surface temperature $T_* \gtrsim 3 \times 10^{4\,\circ}$K, near or in nearly every nebula; and the ultraviolet photons these stars emit transfer energy to the nebula by photoionization. In nebulae and in practically all astronomical objects, H is by far the most abundant element, and photo-ionization of H is thus the main energy input mechanism. Photons with energy greater than 13.6 eV, the ionization potential of H, are absorbed in this process, and the excess energy of each absorbed photon over the ioniza-tion potential appears as kinetic energy of a newly liberated photoelectron. Collisions between electrons, and between electrons and ions, distribute this energy and maintain a Maxwellian velocity distribution with temperature T in the range $5000\,^\circ$K $< T < 20{,}000\,^\circ$K in typical nebulae. Collisions between thermal electrons and ions excite the low-lying energy levels of the ions. Downward radiation transitions from these excited levels have very small transition probabilities, but at the low densities ($N_e \lesssim 10^4$ cm^{-3}) of typical nebulae, collisional de-excitation is even less probable, so almost every excitation leads to emission of a photon, and the nebulae thus emit a forbidden-line spectrum that is quite difficult to excite under terrestrial laboratory conditions.

Thermal electrons are recaptured by the ions, and the degree of ionization at each point in the nebula is fixed by the equilibrium between photoioniza-tion and recapture. In nebulae in which the central star has an especially high temperature, T_*, the radiation field has a correspondingly high number of high-energy photons, and the nebular ionization is therefore high. In such nebulae collisionally excited lines up to [Ne V] and [Fe VII] may be ob-served, but it is important to realize that the high ionization results from the high energy of the photons emitted by the star, and does not necessarily indicate a high nebular temperature T, defined by the kinetic energy of the free electrons.

In the recombination process, recaptures occur to excited levels, and excited atoms thus formed then decay to lower and lower levels by radiative transitions, eventually ending in the ground level. In this process, line photons are emitted, and this is the origin of the observed H I Balmer- and Paschen-line spectra observed in all gaseous nebulae. Note that the recombi-nation of H$^+$ gives rise to excited atoms of H^0 and thus leads to the emission of the H I spectrum. Likewise, He$^+$ recombines and emits the He I spectrum, while in the most highly ionized regions, He^{++} recombines and emits the He II spectrum, the strongest line in the ordinary observed region being $\lambda 4686$. Much weaker recombination lines of C II, C III, C IV, and so on, are also emitted, but in fact, as we shall see, the main excitation process responsible for the observed strength of these lines is resonance fluorescence by photons, which is much less effective for H and He lines because of the greater optical depths in the resonance lines of these more abundant elements.

In addition to the bright-line and continuous spectra emitted by atomic processes, many nebulae also have an infrared continuous spectrum emitted by dust particles heated to a temperature of order 100° by radiation derived originally from the central star.

Gaseous nebulae may be classified into two main types, *diffuse nebulae* or *H II regions,* and *planetary nebulae.* Though the physical processes in both types are quite similar, the origin, mass, evolution, and age of typical members of the two groups are quite different, and thus for some purposes it is convenient to discuss them separately. In addition, a much rarer class of objects, supernova remnants, differs rather greatly from both diffuse and planetary nebulae. In the remainder of this introduction, we shall briefly examine each of these types of objects.

1.4 Diffuse Nebulae

Diffuse nebulae or H II regions are regions of interstellar gas in which the exciting star or stars are O- or early B-type stars of Population I. In many cases there are several exciting stars, often a multiple star, or a galactic cluster of which the hottest two or three stars are the main sources of ionizing radiation. These hot luminous stars undoubtedly formed fairly recently from interstellar matter that in many cases would otherwise be part of the same nebula they now ionize and thus illuminate. The effective temperatures of the stars are in the range $3 \times 10^{4}°K < T_* < 5 \times 10^{4}°K$, and throughout the nebula, H is ionized. He is singly ionized, and other elements are mostly singly or doubly ionized. Typical densities in the ionized part of the nebula are of order 10 or 10^2 cm^{-3}, ranging as high as 10^4 cm^{-3}, although undoubtedly small denser regions exist close to or even below the limit of resolvability. In many nebulae dense neutral condensations are scattered through the ionized volume. Internal motions occur in the gas with velocities of order 10 km sec^{-1}, approximately the isothermal sound speed. Bright rims, knots, condensations, and so on, are apparent to the limit of resolution. There is a tendency for the hot ionized gas to expand into the cooler surrounding neutral gas, thus decreasing the density within the nebula and increasing the ionized volume. The outer edge of the nebula is surrounded by ionization fronts running out into the neutral gas.

The spectra of these "H II regions," as they are often called (because they contain mostly H$^+$), are strong in H I recombination lines and [N II] and [O II] collisionally excited lines, while the strengths of [O III] and [Ne III] lines

FIGURE 1.1 NGC 6611, a bright H II region in which the exciting O stars are members of a star cluster. Note the dark condensations, bright rims, and other fine structure. The area shown is approximately one-half degree in diameter, or 20 pc at the distance of the nebula. Original plate taken in Hα and [N II] λλ6548, 6583 with the 200-inch Hale telescope. (*Hale Observatories photograph.*)

are variable, being stronger in the nebulae with higher central-star temperatures.

These H II regions are observed not only in our Galaxy but also in other
nearby galaxies. The brightest H II regions can easily be seen on almost
any large-scale photograph, but plates taken in a narrow wavelength band
in the red, including Hα and the [N II] lines, are especially effective in
showing faint and often heavily reddened H II regions in other galaxies.
The H II regions are strongly concentrated to the spiral arms, and indeed
are the best objects for tracing the structure of spiral arms in distant galaxies.
Radial-velocity measurements of H II regions then give information on the
kinematics of Population I objects in our own and other galaxies. Typical
masses of observed H II regions are of order 10^2 to 10^4 $M_\odot$, with the lower
limit, of course, a strong function of the sensitivity of the observational
method used.

1.5   Planetary Nebulae

Planetary nebulae are isolated nebulae, often (but not always) possessing
a fair degree of bilateral symmetry, that are actually shells of gas that have
been lost in the fairly recent past by their central stars. The name "planetary"
is purely historical and refers to the fact that some of the bright planetaries
appear as small, disklike objects in small telescopes. The central stars of
planetary nebulae are old stars, typically considerably hotter than galactic
O stars ($T_* \approx 5 \times 10^{4\circ}$K to $3 \times 10^{5\circ}$K) and often less luminous ($M_V = -3$
to $+5$). The stars are in fact rapidly evolving toward the white-dwarf stage,
and the shells are expanding with velocities of order of several times the
velocity of sound (25 km sec^{-1} is a typical expansion velocity). However,
because they are decreasing in density, their emission is decreasing, and on
a cosmic time scale they rapidly become unobservable, with mean lifetimes
as planetary nebulae of a few times 10^4 yr.

As a consequence of the higher stellar temperatures of their exciting stars,
typical planetary nebulae have a higher degree of ionization than do H II
regions, often including large amounts of He^{++}. Their spectra thus include
not only the H I and He I recombination lines, but also, in many cases, the
He II lines, while the collisionally excited lines of [O III] and [Ne III] are
characteristically stronger than those in diffuse nebulae, and [Ne V] is often
strong. There is a wide range in planetary-nebula central stars, however,
and the lower-ionization planetaries have spectra that are quite similar to
those of H II regions.

The space distribution and kinematic properties of planetary nebulae
indicate that, on the cosmic time scale, they are fairly old objects, usually
classified as old Disk Population or old Population I objects. This indicates

8

FIGURE 1.2

NGC 7293, a nearby, large, low-surface-brightness planetary nebula. Note the fine structure, including the long narrow radial filaments in the central "hole" of the nebula, which point to the exciting star. The diameter of the nebula is approximately one quarter of a degree, or 0.5 pc at the distance of the nebula. Original plate taken in Hα and [N II] λλ6548, 6583 with the 200-inch Hale telescope. (*Hale Observatories photograph.*)

that the bulk of the planetaries we now see, though relatively young as planetary nebulae, are actually near-terminal stages in the evolution of quite old stars.

Typical densities in observed planetary nebulae range from 10^4 cm^{-3} down to 10^2 cm^{-3}, and typical masses are of order 0.1 $M_\odot$ to 1.0 $M_\odot$. A few bright planetaries have been observed in other nearby galaxies, specifically the Magellanic Clouds and M 31, but as their luminosities are much smaller than the luminosities of the brightest H II regions, they are difficult to study in detail.

1.6 Supernova Remnants

A few emission nebulae are known to be supernova remnants. The Crab Nebula (NGC 1952), the remnant of the supernova of 1054 A.D., is the best-known example, and small bits of scattered nebulosity are the observable remnants of the much more heavily reddened objects, Tycho's supernova of 1577 and Kepler's supernova of 1604. All three of these supernova remnants have strong nonthermal radio spectra, and several other filamentary nebulae with appearances quite unlike typical diffuse or planetary nebulae have been identified as older supernova remnants by the fact that they have similar nonthermal radio spectra. Two of the best-known examples are the Cygnus Loop (NGC 6960-6992-6995) and IC 443. In the Crab Nebula, the nonthermal synchrotron spectrum observed in the radio-frequency region extends into the optical region, and extrapolation to the ultraviolet region indicates that it is probably the source for the photons responsible for ionizing the nebula. However, in the other supernova remnants no photoionization source is seen, and much of the energy is instead provided by the conversion of kinetic energy of motion into heat. In other words, the fast-moving filaments collide with ambient interstellar gas, and the energy thus released provides ionization and thermal energy, which later is partly radiated as recombination- and collisional-line radiation. Thus these supernova remnants are objects in which *collisional ionization* occurs, rather than photoionization, but the reader should carefully note that in all the nebulae, *collisional excitation* is caused by the thermal electrons that are energized either by photoionization or collisional ionization.

Though nearly all the ideas used in interpreting H II regions and planetary nebulae also apply to supernova remnants, the latter are sufficiently different in detail that they will not be discussed further in this book, which is thus devoted entirely to the study of photoionization nebulae.

References

Every introductory textbook on astronomy contains an elementary general description of gaseous nebulae. The following books and papers are good general references for the entire subject.

Middlehurst, B. M., and Aller, L. H., eds. 1960. *Nebulae and Interstellar Matter*. Chicago: University of Chicago Press.
Terzian, Y. ed. 1968. *Interstellar Ionized Hydrogen*. New York: Benjamin.
Aller, L. H. 1956. *Gaseous Nebulae*. London: Chapman-Hall.
Spitzer, L. 1964. *Diffuse Matter in Space*. New York: Wiley.
Seaton, M. J. 1960. *Reports Progress Phys.* **23,** 313.
Osterbrock, D. E. 1964. *Ann. Rev. Astr. and Astrophys.* **2,** 95.
Osterbrock, D. E. 1967. *P. A. S. P.* **79,** 523.
Lynds, B. T. 1965. *Ap. J. Supp.* **12,** 163.

The last reference is a catalogue of bright nebulae, including emission and reflection nebulae, identified on the National Geographic–Palomar Observatory Sky Survey, taken with the 48-inch Schmidt. This paper contains references to several earlier catalogues.

2

Photoionization Equilibrium

2.1   Introduction

Emission nebulae result from the photoionization of a diffuse gas cloud by
ultraviolet photons from a hot "exciting" star or from a cluster of exciting
stars. The ionization equilibrium at each point in the nebula is fixed by
the balance between photoionizations and recombinations of electrons with
the ions. Since hydrogen is the most abundant element, we can get a first
idealized approximation to the structure of a nebula by considering a pure
H cloud surrounding a single hot star. The ionization equilibrium equation
is:

$$N_{\mathrm{H}^0} \int_{\nu_0}^{\infty} \frac{4\pi J_\nu}{h\nu} a_\nu(\mathrm{H}) \, d\nu = N_e N_p \alpha(\mathrm{H}^0, T), \qquad (2.1)$$

where J_ν is the mean intensity of radiation (in energy units per unit area
per unit time per unit solid angle per unit frequency interval) at the point.
Thus $4\pi J_\nu/h\nu$ is the number of incident photons per unit area per unit time
per unit frequency interval, and $a_\nu(\mathrm{H})$ is the ionization cross section for H

by photons with energy $h\nu$ (above the threshold $h\nu_0$); the integral therefore represents the number of photoionizations per H atom per unit time. N_{H^0}, N_e, and N_p are the neutral atom, electron, and proton densities per unit volume, and $\alpha(H^0, T)$ is the recombination coefficient, so the right-hand side of the equation gives the number of recombinations per unit volume per unit time.

To a first approximation, the mean intensity is simply the radiation emitted by the star reduced by the inverse-square effect of geometrical dilution. Thus

$$4\pi J_\nu = \frac{R^2}{r^2}\,\pi F_\nu(0) = \frac{L_\nu}{4\pi r^2}, \tag{2.2}$$

where R is the radius of the star, $\pi F_\nu(0)$ is the flux at the surface of the star, r is the distance from the star to the point in question, and L_ν is the luminosity of the star per unit frequency interval.

At a typical point in a nebula, the ultraviolet radiation field is so intense that the H is almost completely ionized. Consider, for example, a point in an H II region, with density 10 H atoms and ions per cm³, 5 pc from a central O6 star with $T_* = 40{,}000°K$. We shall examine the numerical values of all the other variables later, but for the moment we can adopt the following very rough values:

$$\int_{\nu_0}^{\infty} \frac{L_\nu\,d\nu}{h\nu} = 5 \times 10^{48} \text{ photons sec}^{-1};$$

$$a_\nu(H) \approx 6 \times 10^{-18} \text{ cm}^2;$$

$$\int_{\nu_0}^{\infty} \frac{4\pi J_\nu}{h\nu}\,a_\nu(H)\,d\nu \approx 10^{-8} \text{ sec}^{-1};$$

$$\alpha(H^0, T) \approx 4 \times 10^{-13} \text{ cm}^3 \text{ sec}^{-1}.$$

Substituting these values and taking ξ as the fraction of neutral H, that is, $N_e = N_p = (1 - \xi)N_H$ and $N_{H^0} = \xi N_H$, where $N_H = 10$ cm^{-3} is the density of H, we find $\xi \approx 4 \times 10^{-4}$, that is, H is very nearly completely ionized.

On the other hand, a finite source of ultraviolet photons cannot ionize an infinite volume, and therefore, if the star is in a sufficiently large gas cloud, there must be an outer edge to the ionized material. The thickness of this transition zone between ionized and neutral gas, since it is due to absorption, is approximately one mean free path of an ionizing photon. Using the same parameters as used previously and taking $\xi = 0.5$, we find the thickness

$$d \approx \frac{1}{N_{H^0}a_\nu} \approx 0.01 \text{ pc},$$

or much smaller than the radius of the ionized nebula. Thus we have the picture of a nearly completely ionized "Strömgren sphere" or H II region, separated by a thin transition region from an outer neutral gas cloud or H I region. In the rest of this chapter we shall explore this ionization structure in detail.

First we will examine the photoionization cross section and the recombination coefficients for H, and then use this information to calculate the structure of hypothetical pure H regions. Next we will consider the photoionization cross section and recombination coefficients for He, the next most abundant element after H, and then calculate more realistic models of H II regions, in which both H and He are taken into account. Finally, we extend our analysis to other less abundant heavy elements, which, in most cases, do not strongly affect the ionization structure of the nebula, but which are quite important in the thermal balance to be discussed in the next chapter.

2.2 Photoionization and Recombination of Hydrogen

Figure 2.1 is an energy-level diagram of H, with the individual terms marked with their quantum numbers n (principal quantum number) and L (angular momentum quantum number), and with $S, P, D, F \ldots$ standing for $L = 0$, 1, 2, 3 ... in the conventional notation. Permitted transitions (which, for one-electron systems, must satisfy the section rule $\Delta L = \pm 1$) are marked by solid lines in the figure. The transition probabilities $A_{nL,n'L'}$ of these lines are of order 10^4 to $10^8 \ \text{sec}^{-1}$, and the corresponding mean lifetimes of the excited levels,

$$\tau_{nL} = \frac{1}{\sum_{n'<n} A_{nL,n'L'}}, \qquad (2.3)$$

are therefore of order 10^{-4} to 10^{-8} sec. The only exception is the $2\ ^2S$ level, from which there are no allowed one-photon downward transitions. However, the transition $2\ ^2S \rightarrow 1\ ^2S$ does occur with the emission of two photons, and the probability of this process is $A_{2\ ^2S,\ 1\ ^2S} = 8.23\ \text{sec}^{-1}$, corresponding to a mean lifetime for the $2\ ^2S$ level of 0.12 sec. Even this lifetime is quite short compared with the mean lifetime of an H atom against photoionization, which has been estimated previously as 10^8 sec for the $1\ ^2S$ level, and is of the same order of magnitude for the excited levels. Thus, to a very good approximation, we may consider that very nearly all the H^0 is in the $1\ ^2S$ level, and that photoionization from this level is balanced by recombination to all levels, recombination to any excited level being followed very quickly by radiative transitions downward, leading ultimately to the ground level. This is the basic approximation that greatly simplifies calculations of physical conditions in gaseous nebulae.

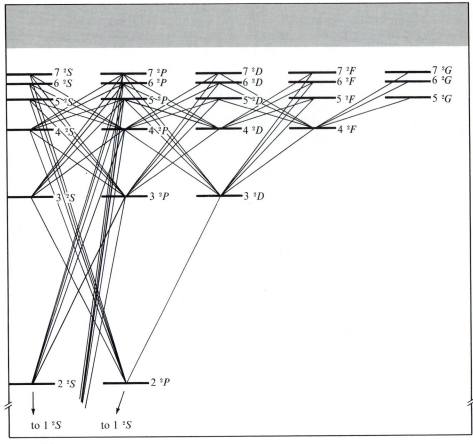

FIGURE 2.1
Partial energy-level diagram of H I, limited to $n \leqslant 7$ and $L \leqslant G$. Permitted radiative transitions to levels $n \leqslant 4$ are indicated by solid lines.

The photoionization cross section for the $1\,^2S$ level of H^0, or, in general, of a hydrogenic ion with nuclear charge Z, may be written in the form

$$a_\nu(Z) = \frac{A_0}{Z^2}\left(\frac{\nu_1}{\nu}\right)^4 \frac{e^{4-(4\tan^{-1}\epsilon)/\epsilon}}{1 - e^{-2\pi/\epsilon}} \qquad \text{for } \nu \geqslant \nu_1, \tag{2.4}$$

where

$$A_0 = \frac{2^8\pi}{3e^4}\left(\frac{1}{137.3}\right)\pi a_0^2 = 6.30 \times 10^{-18}\ \text{cm}^2,$$

$$\epsilon = \sqrt{\frac{\nu}{\nu_1} - 1},$$

and

$$hv_1 = Z^2\, hv_0 = 13.60\, Z^2 \text{ eV}$$

is the threshold energy. This cross section is plotted in Figure 2.2, which shows that it drops off rapidly with energy, approximately as v^{-3} not too far above the threshold, which, for H, is at $v_0 = 3.29 \times 10^{15} \text{ sec}^{-1}$ or $\lambda_0 = 912$ A, so that the higher-energy photons, on the average, penetrate further into neutral gas before they are absorbed.

The electrons produced by photoionization have an initial distribution of energies that depends on $J_v a_v / hv$. However, the cross section for elastic-scattering collisions between electrons is quite large, of order $4\pi(e^2/mv^2)^2 \approx 10^{-13} \text{ cm}^2$, and these collisions tend to set up a Maxwell-Boltzmann energy distribution. The recombination cross section, and all the other cross

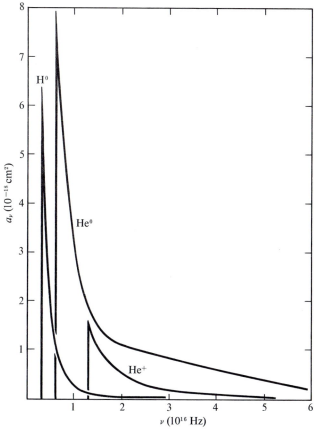

FIGURE 2.2
Photoionization absorption cross sections of H^0, He^0, and H^+.

sections involved in the nebulae, are so much smaller that, to a very good approximation, the electron distribution function is Maxwellian, and therefore all atomic processes occur at rates fixed by the local temperature defined by this Maxwellian. Therefore, the recombination coefficient to a particular level $n\,^2L$ may be written

$$\alpha_{n\,^2L}(H^0, T) = \int_0^\infty v\sigma_{nL}(H^0, v)f(v)\,dv, \tag{2.5}$$

where

$$f(v) = \frac{4}{\sqrt{\pi}} \left(\frac{m}{2kT}\right)^{3/2} v^2 e^{-mv^2/2kT} \tag{2.6}$$

is the Maxwell-Boltzmann distribution function for the electrons, and $\sigma_{nL}\,(H^0, v)$ is the recombination cross section to the term $n\,^2L$ in H^0 for electrons with velocity v. These cross sections vary approximately as v^{-2}, and the recombination coefficients therefore vary approximately as $T^{-1/2}$. A selection of numerical values of $\alpha_{n\,^2L}$ is given in Table 2.1. Since the mean electron velocities at the temperatures listed are of order 5×10^7 cm sec^{-1}, it can be seen that the recombination cross sections are of order 10^{-20} cm^2 or 10^{-21} cm^2, much smaller than the geometrical cross section of an H atom.

In the nebular approximation discussed previously, recombination to any level $n\,^2L$ quickly leads through downward radiative transitions to $1\,^2S$, and the total recombination coefficient is the sum over captures to all levels, ordinarily written

$$\alpha_A = \sum_{n,L} \alpha_{n\,^2L}(H^0, T)$$

$$= \sum_n \sum_{L=0}^{n-1} \alpha_{nL}(H^0, T)$$

$$= \sum_n \alpha_n(H^0, T), \tag{2.7}$$

where α_n is thus the recombination coefficient to all the levels with principal quantum number n. Numerical values of α_A are also listed in Table 2.1. A typical recombination time is $\tau_r = 1/N_e\alpha_A \approx 3 \times 10^{12}/N_e$ sec $\approx 10^5/N_e$ yr, and deviations from ionization equilibrium are ordinarily damped out in times of this order of magnitude.

TABLE 2.1
Recombination Coefficients $\alpha_{n\,^2L}$ *for* H

$\alpha_{n\,^2L}$	T		
	5000°	10,000°	20,000°
$\alpha_{1\,^2S}$	2.28×10^{-13}	1.58×10^{-13}	1.08×10^{-13}
$\alpha_{2\,^2S}$	3.37×10^{-14}	2.34×10^{-13}	1.60×10^{-13}
$\alpha_{2\,^2P}$	8.33×10^{-14}	5.35×10^{-13}	3.24×10^{-13}
$\alpha_{3\,^2S}$	1.13×10^{-14}	7.81×10^{-15}	5.29×10^{-15}
$\alpha_{3\,^2P}$	3.17×10^{-14}	2.04×10^{-14}	1.23×10^{-14}
$\alpha_{3\,^2D}$	3.03×10^{-14}	1.73×10^{-14}	9.09×10^{-15}
$\alpha_{4\,^2S}$	5.23×10^{-15}	3.59×10^{-15}	2.40×10^{-15}
$\alpha_{4\,^2P}$	1.51×10^{-14}	9.66×10^{-15}	5.81×10^{-15}
$\alpha_{4\,^2D}$	1.90×10^{-14}	1.08×10^{-14}	5.68×10^{-15}
$\alpha_{4\,^2F}$	1.09×10^{-14}	5.54×10^{-15}	2.56×10^{-15}
$\alpha_{10\,^2S}$	4.33×10^{-16}	2.84×10^{-16}	1.80×10^{-16}
$\alpha_{10\,^2G}$	2.02×10^{-15}	9.28×10^{-16}	3.91×10^{-16}
$\alpha_{10\,^2M}$	2.7×10^{-17}	1.0×10^{-17}	$4. \times 10^{-18}$
α_A	6.82×10^{-13}	4.18×10^{-13}	2.51×10^{-13}
α_B	4.54×10^{-13}	2.60×10^{-13}	1.43×10^{-13}

* In cm^3 sec^{-1}.

2.3 Photoionization of a Pure Hydrogen Nebula

Consider the simple idealized problem of a single star that is a source of ionizing photons in a homogeneous static cloud of H. Only radiation with frequency $\nu \geqslant \nu_0$ is effective in the photoionization of H from the ground level, and the ionization equilibrium equation at each point can be written

$$N_{H^0} \int_{\nu_0}^{\infty} \frac{4\pi J_\nu}{h\nu} a_\nu \, d\nu = N_p N_e \alpha_A(T). \qquad (2.8)$$

The equation of transfer for radiation with $\nu \geqslant \nu_0$ can be written in the form

$$\frac{dI_\nu}{ds} = -N_{H^0} a_\nu I_\nu + j_\nu, \qquad (2.9)$$

where I_ν is the specific intensity of radiation and j_ν is the local emission coefficient (in energy units per unit volume per unit time per unit solid angle per unit frequency) for ionizing radiation.

It is convenient to divide the radiation field into two parts, a "stellar" part, resulting directly from the input radiation from the star, and a "diffuse" part, resulting from the emission of the ionized gas,

$$I_\nu = I_{\nu s} + I_{\nu d}. \tag{2.10}$$

The stellar radiation decreases outward because of geometrical dilution and absorption, and since its only source is the star, it can be written

$$4\pi J_{\nu s} = \pi F_{\nu s}(r) = \pi F_{\nu s}(R)\frac{R^2 e^{-\tau_\nu}}{r^2}, \tag{2.11}$$

where $\pi F_{\nu s}(r)$ is the standard astronomical notation for the flux of stellar radiation (per unit area per unit time per unit frequency interval) at r, $\pi F_{\nu s}(R)$ is the flux at the radius of the star R, and τ_ν is the radial optical depth at r,

$$\tau_\nu(r) = \int_0^r N_{H^0}(r')a_\nu \, dr', \tag{2.12}$$

which can also be written

$$\tau_\nu(r) = \frac{a_\nu}{a_{\nu_0}}\tau_0(r)$$

in terms of τ_0, the optical depth at the threshold.

The equation of transfer for the diffuse radiation $I_{\nu d}$ is

$$\frac{dI_{\nu d}}{ds} = -N_H a_\nu I_{\nu d} + j_\nu, \tag{2.13}$$

and for $kT \ll h\nu_0$ the only source of ionizing radiation is recaptures of electrons from the continuum to the ground $1\,^2S$ level. The emission coefficient for this radiation is

$$j_\nu(T) = \frac{2h\nu^3}{c^2}\left(\frac{h^2}{2\pi m k T}\right)^{3/2} e^{-h(\nu-\nu_0)/kT} N_p N_e \qquad (\nu > \nu_0), \tag{2.14}$$

which is strongly peaked to $\nu = \nu_0$, the threshold. The total number of photons generated by recombinations to the ground level is given by the recombination coefficient

$$4\pi \int_{\nu_0}^\infty \frac{j_\nu}{h\nu}\,d\nu = N_p N_e \alpha_1(H^0, T), \tag{2.15}$$

and since $\alpha_1 = \alpha_{1\,2S} < \alpha_A$, the diffuse field $J_{\nu d}$ is smaller than $J_{\nu s}$ on the average, and may be calculated by an iterative procedure. For an optically thin nebula, a good first approximation is to take $J_{\nu d} \approx 0$.

On the other hand, for an optically thick nebula, a good first approximation is based on the fact that no ionizing photons can escape, so that every diffuse radiation-field photon generated in such a nebula is absorbed elsewhere in the nebula

$$4\pi \int \frac{j_\nu}{h\nu} \, dV = 4\pi \int N_{\mathrm{H}^0} \frac{a_\nu J_{\nu d}}{h\nu} \, dV, \tag{2.16}$$

where the integration is over the entire volume of the nebula. The on-the-spot approximation amounts to assuming that a similar relation holds locally:

$$J_{\nu d} = \frac{j_\nu}{N_{\mathrm{H}^0} a_\nu}. \tag{2.17}$$

This, of course, automatically satisfies (2.16), and would be exact if all photons were absorbed very close to the point at which they are generated ("on the spot"). This is not a bad approximation because the diffuse radiation-field photons have $\nu \approx \nu_0$, and therefore have large a_ν and correspondingly small mean free paths before absorption.

Making this on-the-spot approximation and using (2.15), the ionization equation (2.8) becomes

$$\frac{N_{\mathrm{H}^0} R^2}{r^2} \int_{\nu_0}^\infty \frac{\pi F_\nu(R)}{h\nu} a_\nu e^{-\tau_\nu} \, d\nu = N_p N_e \alpha_B(\mathrm{H}^0, T), \tag{2.18}$$

where

$$\alpha_B(\mathrm{H}^0, T) = \alpha_A(\mathrm{H}^0, T) - \alpha_1(\mathrm{H}^0, T)$$

$$= \sum_2^\infty \alpha_n(\mathrm{H}^0, T).$$

The physical meaning is that in optically thick nebulae, the ionizations caused by stellar radiation-field photons are balanced by recombinations to excited levels of H, while recombinations to the ground level generate ionizing photons that are absorbed elsewhere in the nebula but have no net effect on the overall ionization balance.

For any stellar input spectrum $\pi F_\nu(R)$, the integral on the left-hand side of (2.18) can be tabulated as a known function of τ_0, since a_ν and τ_ν are known functions of ν. Thus, for any assumed density distribution

$$N_{\mathrm{H}}(r) = N_{\mathrm{H}^0} + N_p$$

and temperature distribution $T(r)$, equations (2.18) and (2.12) can be integrated outward to find $N_{H^0}(r)$ and $N_p(r) = N_e(r)$. Two calculated models for homogeneous nebulae with constant density $N_H = 10$ H atoms plus ions cm^{-3} and constant temperature $T = 7500°K$ are listed in Table 2.2 and graphed in Figure 2.3. For one of these ionization models the assumed $\pi F_\nu(R)$ is a black-body spectrum with $T_* = 40,000°K$, chosen to represent approximately an O6 main-sequence star, while for the other, the $\pi F_\nu(R)$ is a computed model stellar atmosphere with $T_* = 37,450°K$. The table and graph clearly show the expected nearly complete ionization out to a critical radius r_1, at which the ionization drops off abruptly to nearly zero. The central ionized zone is often referred to as an "H II region" (though "H$^+$ region" is a better name), and it is surrounded by an outer neutral H^0 region.

The radius r_1 can be found from (2.18), substituting from (2.12)

$$\frac{d\tau_\nu}{dr} = N_{H^0}a_\nu$$

and integrating over r:

$$R^2 \int_{\nu_0}^{\infty} \frac{\pi F_\nu(R)}{h\nu} d\nu \int_0^\infty d(-e^{-\tau_\nu}) = \int_0^\infty N_p N_e \alpha_B r^2 dr$$

$$= R^2 \int_{\nu_0}^{\infty} \frac{\pi F_\nu(R)}{h\nu} d\nu.$$

TABLE 2.2
Calculated Ionization Distributions for Model H II Regions

r (pc)	$T_* = 4 \times 10^4$ Black-body model		$T_* = 3.74 \times 10^4$ Model stellar atmosphere	
	$\dfrac{N_p}{N_p + N_{H^0}}$	$\dfrac{N_{H^0}}{N_p + N_{H^0}}$	$\dfrac{N_p}{N_p + N_{H^0}}$	$\dfrac{N_H}{N_p + N_{H^0}}$
0.1	1.0	4.5×10^{-7}	1.0	4.5×10^{-7}
1.2	1.0	2.8×10^{-5}	1.0	2.9×10^{-5}
2.2	0.9999	1.0×10^{-4}	0.9999	1.0×10^{-4}
3.3	0.9997	2.5×10^{-4}	0.9997	2.5×10^{-4}
4.4	0.9995	4.4×10^{-4}	0.9995	4.5×10^{-4}
5.5	0.9992	8.0×10^{-4}	0.9992	8.1×10^{-4}
6.7	0.9985	1.5×10^{-3}	0.9985	1.5×10^{-3}
7.7	0.9973	2.7×10^{-3}	0.9973	2.7×10^{-3}
8.8	0.9921	7.9×10^{-3}	0.9924	7.6×10^{-3}
9.4	0.977	2.3×10^{-2}	0.979	2.1×10^{-2}
9.7	0.935	6.5×10^{-2}	0.940	6.0×10^{-2}
9.9	0.838	1.6×10^{-1}	0.842	1.6×10^{-1}
10.0	0.000	1.0	0.000	1.0

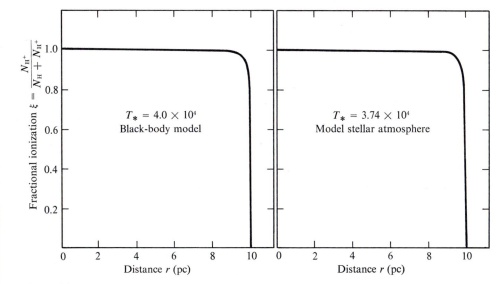

FIGURE 2.3
Ionization structure of two homogeneous pure-H model H II regions.

Using the result that the ionization is nearly complete ($N_p = N_e \approx N_{\mathrm{H}}$) within r_1, and nearly zero ($N_p = N_e \approx 0$) outside r_1, this becomes

$$4\pi R^2 \int_{\nu_0}^{\infty} \frac{\pi F_\nu}{h\nu}\, d\nu = \int_{\nu_0}^{\infty} \frac{L_\nu}{h\nu}\, d\nu$$

$$= Q(\mathrm{H}^0) = \frac{4\pi}{3} r_1^3 N_{\mathrm{H}}^2 \alpha_B. \qquad (2.19)$$

Here $4\pi R^2 \pi F_\nu(R) = L_\nu$ is the luminosity of the star at frequency ν (in energy units per unit time per unit frequency interval), and the physical meaning of (2.19) is that the total number of ionizing photons emitted by the star just balances the total number of recombinations to excited levels within the ionized volume $4\pi r_1^3/3$, often called the Strömgren sphere. Numerical values of radii calculated by using the model stellar atmospheres discussed in Chapter 5 are given in Table 2.3.

2.4 Photoionization of a Nebula Containing Hydrogen and Helium

The next most abundant element after H is He, whose relative abundance (by number) is of order 10 percent, and a much better approximation to the ionization structure of an actual nebula is provided by taking both these

TABLE 2.3
Calculated Radii of Strömgren Spheres

Spectral type	M_V	$T_*(°K)$	Log $Q(H^0)$ (photons/sec)	Log $N_e N_p r_1^3$ (N in cm^{-3}; r_1 in pc)	r_1 ($N_e = N_p$ $= 1$ cm^{-3})
O5	−5.6	48,000	49.67	6.07	108
O6	−5.5	40,000	49.23	5.63	74
O7	−5.4	35,000	48.84	5.24	56
O8	−5.2	33,500	48.60	5.00	51
O9	−4.8	32,000	48.24	4.64	34
O9.5	−4.6	31,000	47.95	4.35	29
B0	−4.4	30,000	47.67	4.07	23
B0.5	−4.2	26,200	46.83	3.23	12

NOTE: $T = 7500°K$ assumed for calculating α_B.

elements into account. The ionization potential of He is $h\nu_2 = 24.6$ eV, somewhat higher than H, while the ionization potential of He^+ is 54.4 eV, but since even the hottest O stars emit practically no photons with $h\nu > 54.4$ eV, the possibility of second ionization of He does not exist in ordinary H II regions (though the situation is quite different in planetary nebulae, as we shall see later in this chapter). Thus photons with energy 13.6 eV $< h\nu < 24.6$ eV can ionize H only, while photons with energy $h\nu > 24.6$ eV can ionize both H and He. As a result, two different types of ionization structure are possible, depending on the spectrum of ionizing radiation and the abundance of He. At one extreme, if the spectrum is concentrated to frequencies just above 13.6 eV and contains only a few photons with $h\nu > 24.6$ eV, then the photons with energy 13.6 eV $< h\nu < 24.6$ eV keep the H ionized, and the photons with $h\nu > 24.6$ eV are all absorbed by He. The ionization structure thus consists of a small central H^+, He^+ zone surrounded by a larger H^+, He^0 region. At the other extreme, if the input spectrum contains a large fraction of photons with $h\nu > 24.6$ eV, then these photons dominate the ionization of both H and He, the outer boundaries of both ionized zones coincide, and there is a single H^+, He^+ region.

The He^0 photoionization cross section $a_\nu(He^0)$ is plotted in Figure 2.2, along with $a_\nu(H^0)$ and $a_\nu(He^+)$ calculated from equation (2.4). The total recombination coefficients for He to configurations $L \geqslant 2$ are, to a good approximation, the same as for H, since these levels are hydrogenlike, but because He is a two-electron system, it has separate singlet and triplet levels and

$$
\left.
\begin{aligned}
\alpha_{n\,^1L}(He^0, T) &\approx \frac{1}{4}\alpha_{n\,^2L}(H^0, T) \\
\alpha_{n\,^3L}(He^0, T) &\approx \frac{3}{4}\alpha_{n\,^2L}(H^0, T)
\end{aligned}
\right\} L \geqslant 2,
\tag{2.20}
$$

while for the P and particularly the S terms there are sizable differences between the He and H recombination coefficients. Representative numerical values of the recombination coefficients are included in Table 2.4.

The ionization equations for H and He are coupled by the radiation field with $h\nu > 24.6$ eV, and are straightforward to write down in the on-the-spot approximation, though complicated in detail. First of all, the photons emitted in recombinations to the ground level of He can ionize either H or He, since these photons are emitted with energies just above $h\nu_2 = 24.6$ eV, and the fraction absorbed by H is

$$y = \frac{N_{\mathrm{H}^0}a_{\nu_2}(\mathrm{H}^0)}{N_{\mathrm{H}^0}a_{\nu_2}(\mathrm{H}^0) + N_{\mathrm{He}^0}a_{\nu_2}(\mathrm{He}^0)}, \qquad (2.21)$$

while the remaining fraction $1 - y$ is absorbed by He. Secondly, following recombination to excited levels of He, various photons are emitted that ionize H. Of the recombinations to excited levels of He, approximately three-fourths are to the triplet levels and approximately one-fourth are to the

TABLE 2.4
He *Recombination Coefficients**

	T		
	5000°K	10,000°K	20,000°K
$\alpha(\mathrm{He}^0, 1\,^1S)$	2.23×10^{-13}	1.59×10^{-13}	1.14×10^{-13}
$\alpha(\mathrm{He}^0, 2\,^1S)$	7.64×10^{-15}	5.55×10^{-15}	4.06×10^{-15}
$\alpha(\mathrm{He}^0, 2\,^1P)$	2.11×10^{-14}	1.35×10^{-14}	8.16×10^{-15}
$\alpha(\mathrm{He}^0, 3\,^1S)$	2.23×10^{-15}	1.63×10^{-15}	1.19×10^{-15}
$\alpha(\mathrm{He}^0, 3\,^1P)$	8.92×10^{-15}	5.65×10^{-15}	3.34×10^{-15}
$\alpha(\mathrm{He}^0, 3\,^1D)$	9.23×10^{-15}	5.28×10^{-15}	2.70×10^{-15}
$\alpha_B(\mathrm{He}^0, n\,^1L)$	9.96×10^{-14}	6.27×10^{-14}	3.46×10^{-14}
$\alpha(\mathrm{He}^0, 2\,^3S)$	1.98×10^{-14}	1.46×10^{-14}	1.13×10^{-14}
$\alpha(\mathrm{He}^0, 2\,^3P)$	8.78×10^{-14}	5.68×10^{-14}	3.59×10^{-14}
$\alpha(\mathrm{He}^0, 3\,^3S)$	4.88×10^{-15}	3.73×10^{-15}	2.97×10^{-15}
$\alpha(\mathrm{He}^0, 3\,^3P)$	3.20×10^{-14}	1.95×10^{-14}	1.30×10^{-14}
$\alpha(\mathrm{He}^0, 3\,^3D)$	2.84×10^{-14}	1.30×10^{-14}	8.46×10^{-15}
$\alpha_B(\mathrm{He}^0, n\,^3L)$	3.26×10^{-13}	2.10×10^{-13}	1.29×10^{-13}
$\alpha_B(\mathrm{He}^0)$	4.26×10^{-13}	2.73×10^{-13}	1.55×10^{-13}

* In cm³ sec⁻¹.

singlet levels. All the captures to triplets lead ultimately through downward radiative transitions to $2\,^3S$, which is highly metastable, but which can decay by a one-photon forbidden line at 19.8 eV to $1\,^1S$, with transition probability $A_{2\,^3S,\,1\,^1S} = 1.27 \times 10^{-4}\ \text{sec}^{-1}$. Competing with this mode of depopulation of $2\,^3S$, collisional excitation to the singlet levels $2\,^1S$, and $2\,^1P$ can also occur with fairly high probability, while collisional transitions to $1\,^1S$ or to the continuum are far less probable. Since these collisions involve a spin change, only electrons are effective in causing the excitation, and the transition rate per atom in the $2\,^3S$ level is

$$N_e q_{2\,^3S,\,2\,^1L} = N_e \int_{(1/2)mv^2 = \chi}^{\infty} v\sigma_{2\,^3S,\,2\,^1L}(v)f(v)\,dv, \qquad (2.22)$$

where the $\sigma_{2\,^3S,\,2\,^1L}(v)$ are the electron collision cross sections for these excitation processes, and the χ are their thresholds. These rates are listed in Table 2.5, along with the critical electron density $N_c(2\,^3S)$, defined by

$$N_c(2\,^3S) = \frac{A_{2\,^3S,\,1\,^1S}}{q_{2\,^3S,\,2\,^1S} + q_{2\,^3S,\,2\,^1P}}, \qquad (2.23)$$

at which collisional transitions are equally probable with radiative transitions. In typical H II regions, the electron density $N_e \lesssim 10^2\ \text{cm}^{-3}$, considerably smaller than N_c, so practically all the atoms leave $2\,^3S$ by emission of a 19.8 eV-line photon, while on the other hand, in typical bright planetary nebulae, $N_e \approx 10^4\ \text{cm}^{-3}$, somewhat larger than N_c, and therefore many of the atoms are transferred to $2\,^1S$ or $2\,^1P$ before emitting a line photon.

Of the captures to the singlet-excited levels in He, approximately two-thirds lead ultimately to population of $2\,^1P$, while approximately one-third leads to population of $2\,^1S$. Atoms in $2\,^1P$ decay mostly to $1\,^1S$ with emission of a resonance-line photon at 21.3 eV, but some also decay to $2\,^1S$ (with emission of $2\,^1S - 2\,^1P$ at 2.2 μ) with a relative probability of approximately 10^{-3}. The resonance-line photons are scattered by He^0, and therefore, after approximately 10^3 scatterings, a typical one would, on the average, be

TABLE 2.5
Collisional Excitation Coefficients from He^0 $(2\,^3S)$

T (°K)	$q_{2\,^3S,\,2\,^1S}$ (cm³ sec⁻¹)	$q_{2\,^3S,\,2\,^1P}$ (cm³ sec⁻¹)	N_c (cm⁻³)
6000	1.9×10^{-8}	1.7×10^{-9}	6.2×10^3
8000	2.5×10^{-8}	3.4×10^{-9}	4.5×10^3
10,000	2.9×10^{-8}	5.0×10^{-9}	3.7×10^3
12,000	3.1×10^{-8}	6.5×10^{-9}	3.4×10^3
15,000	3.3×10^{-8}	8.2×10^{-9}	3.1×10^3
20,000	3.3×10^{-8}	10.4×10^{-9}	2.9×10^3

converted to a 2.2 μ line photon and thus populate $2\ ^1S$. However, it is more likely that before a resonance-line photon is scattered this many times, it will photoionize an H atom and be absorbed. He atoms in $2\ ^1S$ decay by two-photon emission (with the sum of the energies 20.7 eV and transition probability 51 sec^{-1}) to $1\ ^1S$. From the distribution of photons in this continuous spectrum, the probability that a photon is produced that can ionize H is 0.56 per radiative decay from He$^0\ 2\ ^1S$.

All these He bound-bound transitions produce photons that ionize H but not He, and they can easily be included in the H ionization equation in the on-the-spot approximation. The total number of recombinations to excited levels of He per unit volume per unit time is $N_{He^+}N_e\alpha_B(He^0, T)$, and of these a fraction p generate ionizing photons that are absorbed on the spot. As shown by the preceding discussion, in the low-density limit $N_e \ll N_c$,

$$p \approx \frac{3}{4} + \frac{1}{4}\left[\frac{2}{3} + \frac{1}{3}(0.56)\right] = 0.96,$$

while in the high-density limit $N_e \gg N_c$,

$$p \approx \left[\frac{3}{4}(0.90) + \frac{1}{4}\cdot\frac{1}{3}\right](0.56) + \left[\frac{3}{4}(0.10) + \frac{1}{4}\cdot\frac{2}{3}\right] = 0.67.$$

Thus, in the on-the-spot approximation, the ionization equations become

$$\frac{N_{H^0}R^2}{r^2}\int_{\nu_0}^{\infty}\frac{\pi F_\nu(R)}{h\nu}a_\nu(H^0)e^{-\tau_\nu}\,d\nu + yN_{He^+}N_e\alpha_1(He^0, T)$$

$$+ p\,N_{He^+}N_e\alpha_B(He^0, T) = N_pN_e\alpha_B(H^0, T); \quad (2.24)$$

$$\frac{N_{He^0}R^2}{r^2}\int_{\nu_2}^{\infty}\frac{\pi F_\nu(R)}{h\nu}a_\nu(He^0)e^{-\tau_\nu}\,d\nu + (1-y)N_{He^+}N_e\alpha_1(He^0, T)$$

$$= N_{He^+}N_e\alpha_A(He^0, T), \quad (2.25)$$

with

$$\frac{d\tau_\nu}{dr} = N_{H^0}a_\nu(H^0) \qquad \text{for } \nu_0 < \nu < \nu_2$$

and   (2.26)

$$\frac{d\tau_\nu}{dr} = N_{H^0}a_\nu(H^0) + N_{He^0}a_\nu(He^0) \qquad \text{for } \nu_2 < \nu,$$

and

$$N_e = N_p + N_{He^+}.$$

These equations again can be integrated outward step-by-step, and sample models for a diffuse nebula (with $N_H = 10\ cm^{-3}$, $N_{He}/N_H = 0.15$) excited by an O6 star and by a B0 star are plotted in Figure 2.4. It can be seen that the hotter O6 star excites a coincident H^+, He^+ zone, while the cooler B0 star has an inner H^+, He^+ zone and an outer H^+, He^0 zone.

Although the exact size of the He^+ zone can only be found from the integration because of the coupling between the H and He ionization by the radiation with $\nu > \nu_2$, the approximate size can easily be found by

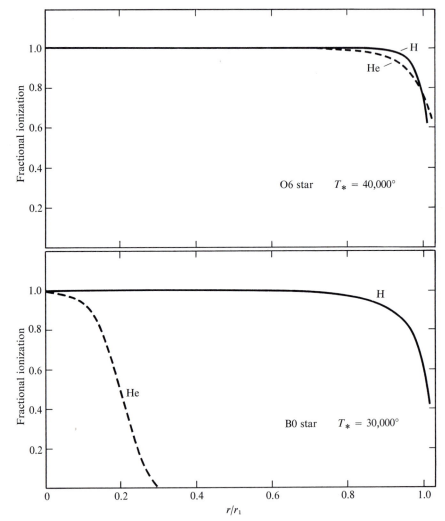

FIGURE 2.4
Ionization structure of two homogeneous H + He model H II regions.

ignoring the absorption of H in the He$^+$ zone. This corresponds to setting $y = 0$ in equation (2.25) and $N_{H^0} = 0$ in the second of equations (2.26), and we then immediately find, in analogy to equation (2.19), that

$$\int_{\nu_2}^{\infty} \frac{L_\nu}{h\nu}\, d\nu = Q(\text{He}^0) = \frac{4\pi}{3} r_2^3 N_{\text{He}^+} N_e \alpha_B(\text{He}^0), \qquad (2.27)$$

where r_2 is the radius of the He$^+$ zone. Furthermore, since, according to the preceding discussion, $p \approx 1$, the absorptions by He do not greatly reduce the number of photons available for ionizing H, and therefore, to a fair approximation,

$$\int_{\nu_0}^{\infty} \frac{L_\nu}{h\nu}\, d\nu = Q(\text{H}^0) = \frac{4\pi}{3} r_1^3 N_{\text{H}^+} N_e \alpha_B(\text{H}^0). \qquad (2.19)$$

If we suppose that the He$^+$ zone is much smaller than the H$^+$ zone, then throughout most of the H$^+$ zone, the electrons come only from ionization of H, while in the He$^+$ zone, the electrons come from ionization of both H and He, and with this simplification,

$$\left(\frac{r_1}{r_2}\right)^3 = \frac{Q(\text{H}^0)}{Q(\text{He}^0)} \frac{N_{\text{He}}}{N_{\text{H}}} \left(1 + \frac{N_{\text{He}}}{N_{\text{H}}}\right) \frac{\alpha_B(\text{He}^0)}{\alpha_B(\text{H}^0)} \qquad \text{if } r_2 < r_1. \qquad (2.28)$$

A plot of r_2/r_1, calculated according to this equation for $N_{\text{He}}/N_{\text{H}} = 0.15$ and $T = 7500°\text{K}$, is shown in Figure 2.5, and it can be seen that, for

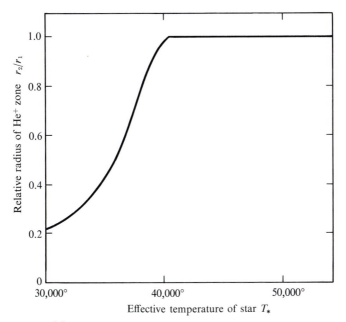

FIGURE 2.5
Relative radius of He$^+$ zone as a function of effective temperature of exciting star.

$T_* \gtrsim 40,000\,°K$, the He^+ and H^+ zones are coincident, while at significantly lower temperatures, the He^+ zone is much smaller. The details of the curve, including the precise temperature at which $r_2/r_1 = 1$, are not significant because of the simplifications made, but the general trends it indicates are correct.

All these models refer to homogeneous, uniform-density H II regions, though of course, it is quite straightforward to modify the equations to include any known or assumed radial dependence of density $N(r)$ or temperature $T(r)$. Actual nebulae, however, have a much more complicated structure with condensations, fronts, density fluctuations, and so on. Some simple but reasonably realistic way of taking this fine-scale structure into account is needed before the ionization structure of nebulae can be said to be completely understood.

2.5   Photoionization of He^+ to He^{++}

Although ordinary Population I O stars do not radiate any appreciable number of photons with $h\nu > 54.4$ eV (with the result that galactic H II regions do not have a He^{++} zone), the situation is quite different for the central stars of planetary nebulae. Many of these stars are much hotter than even the hottest O5 stars and do radiate high-energy photons that produce central He^{++} zones, which are observed by the He II recombination spectra they emit.

The structure of these central He^{++} zones is governed by equations that are very similar to those for pure H^+ zones discussed previously, with the threshold, absorption cross section, and recombination coefficient changed from H^0 ($Z = 1$) to He^+ ($Z = 2$). This He^{++} zone is, of course, also an H^+ zone, and the ionization equations of H^0 and of He^+ are therefore, in principle, coupled, but they are actually separable to a fairly good approximation. This results from the fact that, in the recombination of He^{++} to form He^+, photons are emitted that ionize H^0. Three different mechanisms are involved, namely, recombinations that populate $2\,^2P$, resulting in He II $L\alpha$ emission with $h\nu = 40.7$ eV; recombinations that populate $2\,^2S$, resulting in He II $2\,^2S \to 1\,^2S$ two-photon emission for which $h\nu' + h\nu'' = 40.7$ eV (the spectrum peaks at 20.4 eV, and, on the average, 1.42 ionizing photons are emitted per decay); and recombinations directly to $2\,^2S$ and $2\,^2P$, resulting in He II Balmer continuum emission, which has the same threshold as the Lyman limit of H and therefore emits a continuous spectrum concentrated just above $h\nu_0$. The He II $L\alpha$ photons are scattered by resonance scattering, and therefore diffuse only slowly away from their

TABLE 2.6
Generation Rates of Ionizing Photons in the He^{++} Zone

Generation rate	$T = 1 \times 10^4$	2×10^4
$q(\text{He}^+ \, L\alpha)/\alpha_B(\text{H}^0)$	0.66	0.68
$q(\text{He}^+ \, 2 \text{ photon})/\alpha_B(\text{H}^0)$	0.34	0.40
$q(\text{He}^+ \, Ba \, c)/\alpha_B(\text{H}^0)$	0.27	0.33

NOTE: Numerical values are calculated assuming that
$N(\text{He}^{++})/N(\text{H}^+) = 0.15$.

point of origin before they are absorbed, while the He II Balmer-continuum photons are concentrated close to the H^0 ionization threshold and therefore have a short mean free path. Both these sources thus tend to ionize H^0 in the He^{++} zone, and at a "normal" abundance of He, He/H ≈ 0.15, the number of ionizing photons generated in the He^{++} zone by these two processes is just about sufficient to balance the recombinations of H$^+$ in this zone and thus maintain the ionization of H^0. This is shown in Table 2.6, which lists the ionizing-photon generation rates and the recombination rate for several temperatures. Thus, to a good approximation, the He II $L\alpha$ and Balmer continuum photons are absorbed by and maintain the ionization of H^0 in the He^{++} zone, while the stellar radiation with 13.6 eV $< h\nu <$ 54.4 eV is not significantly absorbed by the H^0 in the He^{++}, H$^+$ zone, and that with $h\nu >$ 54.4 eV is absorbed only by the He$^+$. The He II two-photon continuum is an additional source of ionizing photons for H, and most of these photons escape from the He^{++} zone and therefore must be added to the stellar radiation field with $h\nu <$ 54.4 eV in the He$^+$ zone. Of course, a more accurate calculation may be made, taking into account the detailed frequency dependence of each of the emission processes, but since normally the helium abundance is small, only an approximation to its effects is usually required.

Some sample calculations of the ionization structure of a model planetary nebula with the radiation source a black body at $T_* = 10^5 °$K are shown in Figure 2.6. The sharp outer edge of the He^{++} zone, as well as the even sharper outer edges of the H$^+$ and He$^+$ zones, can be seen in these graphs. There is, of course, an equation that is exactly analogous to (2.19) and (2.27) for the "Strömgren radius" r_3 of the He^{++} zone:

$$\int_{4\nu_0}^{\infty} \frac{L_\nu}{h\nu} \, d\nu = \frac{4\pi}{3} r_3^3 N_{\text{He}^{++}} N_e \alpha_B(\text{He}^+, T). \tag{2.29}$$

Thus stellar temperatures $T_* \gtrsim 10^5 °$K are required for $r_3/r_1 \approx 1$.

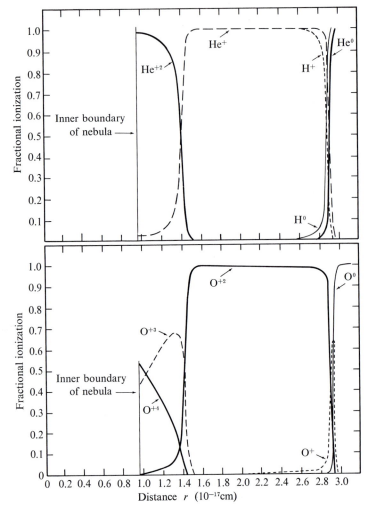

FIGURE 2.6
Ionization structure of H, He (*top*), and O (*bottom*) for a model planetary nebula.

2.6 Further Iterations of the Ionization Structure

As described previously, the on-the-spot approximation may be regarded as the first approximation to the ionization and, as will be described in Chapter 3, to the temperature distribution in the nebula. From these a first approximation to the emission coefficient j_ν may be found throughout the model: from j_ν a first approximation to $I_{\nu d}$ and hence J_ν at each point, and from J_ν a better approximation to the ionization and temperature at each point. This iteration procedure can be repeated as many times as desired (given sufficient computing time) and actually converges quite rapidly, but, except where high accuracy is required, the first (on-the-spot) approximation is usually sufficient.

2.7   Photoionization of Heavy Elements

Finally, let us examine the ionization of the heavy elements, of which O, Ne, C, N, Fe, Si with abundances (by number) of order 10^{-3} to 10^{-4} that of H, are the most abundant. The ionization-equilibrium equation for any two successive stages of ionization i and $i + 1$ of any element X may be written

$$N(X^{+i+1}) \int_{\nu_i}^{\infty} \frac{4\pi J_\nu}{h\nu} a_\nu(X^{+i})\, d\nu = N(X^{+i}) N_e \alpha_G(X^{+i+1}, T), \qquad (2.30)$$

where $N(X^{+i})$ and $N(X^{+i+1})$ are the number densities of the two successive stages of ionization; $a_\nu(X^{+i})$ is the photoionization cross section from the ground level of X^i with the threshold ν_i; and $\alpha_G(X^{+i+1}, T)$ is the recombination coefficient of the ground level of X^{+i+1}, to all levels of X^{+i}. These equations, together with the total number of ions of all stages of ionization,

$$N(X^0) + N(X^{+1}) + N(X^{+2}) + \cdots + N(X^{+n}) = N(X)$$

(presumably known from the abundance of X), completely determine the ionization equilibrium at each point. The mean intensity J_ν, of course, includes both the stellar and diffuse contributions, but the abundances of the heavy elements are so small that their contributions to the diffuse field are negligible, and only the emission by H, He, and He^+ mentioned previously needs to be taken into account.

The required data of numerical values of a_ν and α_G are less readily available for heavy elements, which are many-electron systems, than for H and He. However, approximate calculations, mostly based on Hartree-Fock or close-coupling wave functions, are available for many common ions. For simple ions, photoionization, which is the removal of one outer electron, can lead to only a single level, the ground level of the resulting ion. However, in more complicated ions, photoionization can often lead to any of several levels of the ground configuration of the resulting ion, and correspondingly the photoionization cross section has several thresholds, instead of a single

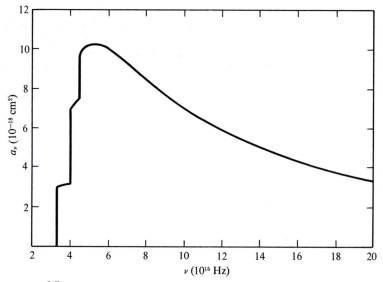

FIGURE 2.7
Absorption cross section of O^0, showing several thresholds resulting from photoionization to several levels of O^+.

threshold as for simpler ions. An example is neutral O, which can be photoionized by the following schemes:

$$O^0(2p^4\,{}^3P) + h\nu \rightarrow \left.\begin{array}{l} O^+(2p^3\,{}^4S) + ks \\[4pt] O^+(2p^3\,{}^4S) + kd \end{array}\right\} \quad h\nu > 13.6\ \mathrm{eV}$$

$$\left.\begin{array}{l} O^+(2p^3\,{}^2D) + ks \\[4pt] O^+(2p^3\,{}^2D) + kd \end{array}\right\} \quad h\nu > 16.9\ \mathrm{eV}$$

$$\left.\begin{array}{l} O^+(2p^3\,{}^2P) + ks \\[4pt] O^+(2p^3\,{}^2P) + kd \end{array}\right\} \quad h\nu > 18.6\ \mathrm{eV}.$$

The calculated photoionization cross section is plotted in Figure 2.7. Note that inner-shell photoionization can also occur; for example, in neutral O,

$$O^0(2s^2\,2p^4\,{}^3P) + h\nu \rightarrow O^+(2s\,2p^4\,{}^4P) + kp\} \quad h\nu > 28.4\ \mathrm{eV}$$

$$O^+(2s\,2p^4\,{}^2D) + kp\} \quad h\nu > 34.0\ \mathrm{eV}$$

$$\cdots,$$

but the thresholds are generally so high that there is little available radiation and they make no contribution to the photoionization in most nebulae. However, they cannot be ignored if the source has a spectrum that does not drop off rapidly at high energies—as may be the situation in quasars or in nebulae excited by X-ray sources.

A good interpolation formula that fits the contribution of each threshold ν_T to the photoionization cross section is

$$a_\nu = a_T\left[\beta\left(\frac{\nu}{\nu_T}\right)^{-s} + (1-\beta)\left(\frac{\nu}{\nu_T}\right)^{-s-1}\right] \qquad \nu > \nu_T, \qquad (2.31)$$

and the total cross section is then the sum of the contributions of the individual thresholds. A list of numerical values of ν_T, a_T, β, and s for common atoms and ions (including H^0, He^0, and He^+) is given in Table 2.7.

The recombination coefficients for complex ions are dominated by recaptures to excited levels, which may, to a good approximation, be taken to be hydrogenlike. Thus, for example, any ion X^{+i} with configuration $1s^2\,2s^2\,2p^n$,

$$\alpha_G(X^{+i}, T) = \alpha_{2p}(X^{+i}, T) + \sum_{n=3}^{\infty} \alpha_n(X^{+i}, T)$$

$$= \alpha_{2p}(X^{+i}, T) + \alpha_C(X^{+i}, T), \qquad (2.32)$$

where we have defined, by analogy with α_A and α_B,

$$\alpha_C(X^{+i}, T) = \sum_{n=3}^{\infty} \alpha_n(X^{+i}, T), \qquad (2.33)$$

and this term is generally considerably larger than $\alpha_{2p}(X^{+i}, T)$.

To a reasonable approximation the outer shells may be taken to be hydrogenlike, with charge Z given by the stage of ionization. In this one-electron approximation, the recombination coefficient depends on T and Z only, so the H results can be scaled to any ion. A list of numerical values of α_A, α_B, α_C, and α_D ($n \geqslant 4$) is given in Table 2.8.

To find the total recombination coefficient according to equation (2.32), it is then necessary to add the first term, representing captures, directly to the ground configuration. This part is far from hydrogenlike, but it can be found from the photoionization cross section, using the Milne relation as explained in Appendix 1. Note that captures to all levels, ground and excited, of the ground configuration should be taken into account. Actually, it is usually sufficient simply to estimate this term or even to ignore it, as it is small in comparison with the term representing the captures to all excited configurations.

Calculations have been made of the ionization of heavy elements in several model H II regions and planetary nebulae. In H II regions, the common elements tend to be most singly ionized in the outer parts of the nebulae, such as O^+ and N^+, though near the central stars there are often fairly considerable amounts of O^{++}, N^{++}, and Ne^{++}. Most planetary nebulae have hotter central stars, and the degree of ionization is correspondingly

TABLE 2.7
Photoionization Cross-section Parameters

Parent	Resulting ion	ν_T (cm^{-1})	a_T (10^{-18} cm^2)	β	s
$H^0(^2S)$	$H^+(^1S)$	1.097×10^5	6.30	1.34	2.99
$He^0(^1S)$	$He^+(^2S)$	1.983×10^5	7.83	1.66	2.05
$He^+(^2S)$	$He^{+2}(^1S)$	4.389×10^5	1.58	1.34	2.99
$C^0(^3P)$	$C^+(^2P)$	9.09×10^4	12.2	3.32	2.0
$C^+(^2P)$	$C^{+2}(^1S)$	1.97×10^5	4.60	1.95	3.0
$C^{+2}(^1S)$	$C^{+3}(^2S)$	3.86×10^5	1.60	2.6	3.0
$C^{+3}(^2S)$	$C^{+4}(^1S)$	5.20×10^5	0.68	1.0	2.0
$N^0(^4S)$	$N^+(^3P)$	1.17×10^5	11.4	4.29	2.0
$N^+(^3P)$	$N^{+2}(^2P)$	2.39×10^5	6.65	2.86	3.0
$N^{+2}(^2P)$	$N^{+3}(^1S)$	3.83×10^5	2.06	1.63	3.0
$N^{+3}(^1S)$	$N^{+4}(^2S)$	6.25×10^5	1.08	2.6	3.0
$N^{+4}(^2S)$	$N^{+5}(^1S)$	7.90×10^5	0.48	1.0	2.0
$O^0(^3P)$	$O^+(^4S)$	1.098×10^5	2.94	2.66	1.0
$O^0(^3P)$	$O^+(^2D)$	1.363×10^5	3.85	4.38	1.5
$O^0(^3P)$	$O^+(^2P)$	1.500×10^5	2.26	4.31	1.5
$O^+(^4S)$	$O^{+2}(^3P)$	2.836×10^5	7.32	3.84	2.5
$O^{+2}(^3P)$	$O^{+3}(^2P)$	4.432×10^5	3.65	2.01	3.0
$O^{+3}(^2P)$	$O^{+4}(^1S)$	6.244×10^5	1.27	0.83	3.0
$O^{+4}(^1S)$	$O^{+5}(^2S)$	9.187×10^5	0.78	2.6	3.0
$O^{+5}(^2S)$	$O^{+6}(^1S)$	1.114×10^6	0.36	1.0	2.1
$Ne^0(^1S)$	$Ne^+(^2P)$	1.739×10^5	5.35	3.77	1.0
$Ne^+(^2P)$	$Ne^{+2}(^3P)$	3.314×10^5	4.16	2.72	1.5
$Ne^+(^2P)$	$Ne^{+2}(^1D)$	3.572×10^5	2.71	2.15	1.5
$Ne^+(^2P)$	$Ne^{+2}(^1S)$	3.871×10^5	0.52	2.13	1.5
$Ne^{+2}(^3P)$	$Ne^{+3}(^4S)$	5.141×10^5	1.80	2.28	2.0
$Ne^{+2}(^3P)$	$Ne^{+3}(^2D)$	5.551×10^5	2.50	2.35	2.5
$Ne^{+2}(^3P)$	$Ne^{+3}(^2P)$	5.763×10^5	1.48	2.23	2.5
$Ne^{+3}(^4S)$	$Ne^{+4}(^3P)$	7.839×10^5	3.11	1.96	3.0
$Ne^{+4}(^3P)$	$Ne^{+5}(^2P)$	1.020×10^6	1.40	1.47	3.0
$Ne^{+5}(^2P)$	$Ne^{+6}(^1S)$	1.274×10^6	0.49	1.15	3.0

TABLE 2.8
Recombination Coefficients for H-like Ions

			T		
	1250°K	2500°K	5000°K	10,000°K	20,000°K
$\alpha_A = \sum_1^\infty \alpha_n$	1.74×10^{-12}	1.10×10^{-12}	6.82×10^{-13}	4.18×10^{-13}	2.51×10^{-13}
$\alpha_B = \sum_2^\infty \alpha_n$	1.28×10^{-12}	7.72×10^{-13}	4.54×10^{-13}	2.60×10^{-13}	1.43×10^{-13}
$\alpha_C = \sum_3^\infty \alpha_n$	1.03×10^{-12}	5.99×10^{-13}	3.37×10^{-13}	1.83×10^{-13}	9.50×10^{-14}
$\alpha_D = \sum_4^\infty \alpha_n$	8.65×10^{-13}	4.86×10^{-13}	2.64×10^{-13}	1.37×10^{-13}	6.83×10^{-14}

NOTE: In this table, $Z = 1$; for other values of Z, $\alpha(Z,T) = Z\alpha(1,T/Z^2)$. Coefficients are measured in cm^3 sec^{-1}.

higher. This is shown in Figure 2.6, where the ionization of O is plotted for a calculated model planetary nebula. Note that in this figure the outer edge of the He^{++} zone is also the outer edge of the O^{+3} zone and the inner edge of the O^{++} zone, since O^{++} has an ionization potential 54.9 eV, nearly the same as He$^+$.

Again, in actual nebulae, density condensations play an important role in complicating the ionization structure; these simplified models do, however, give an overall picture of the ionization.

2.8   Charge-Exchange Reactions

One other atomic process is important in determining the ionization equilibrium of particular light elements, especially near the outer boundaries of radiation-bounded nebulae. This process is charge exchange in two-body reactions with hydrogen. As an example, consider neutral oxygen, which has the charge-exchange reaction with a proton

$$O^0(^3P) + H^+ \rightarrow O^+(^4S) + H^0(^2S). \tag{2.34}$$

This reaction results in the conversion of an originally neutral O atom into an O$^+$ ion, and this is an ionization process for O. There is an attractive polarization force between O° and H+; in addition, the ionization potentials of O and H are very nearly the same, so that the reaction is very nearly a resonance process, and for both these reasons the cross section for this charge-exchange reaction is relatively large. The reaction rate per unit volume per unit time for the reaction can be written

$$N_O \, N_p \, \delta(T), \tag{2.35}$$

where $\delta(T)$ is expressed in terms of the reaction cross section $\sigma(v)$ by an integral analogous to equation (2.5),

$$\delta(T) = \int_0^\infty v \, \sigma(v) f(v) \, dv. \tag{2.36}$$

Here it should be noted that $f(v)$ is the Maxwell-Boltzmann distribution function for the relative velocity v in the OH+ center-of-mass system, and thus involves their reduced mass. A selection of computed values of $\delta(T)$ is given in Table 2.9; for instance, in an H II region with $N_{H+} = 10 \text{ cm}^{-3}$, the ionization rate per O atom per unit time is about 10^{-8} sec^{-1}, comparable with the photoionization rate for the typical conditions adopted in Section 2.1. Likewise, the rate for the inverse reaction

$$O^+(^4S) + H^0(^2S) \to O^0(^3P) + H^+ \tag{2.37}$$

can be written

$$N_{O^+} \, N_H \, \delta'(T). \tag{2.38}$$

Numerical values of $\delta'(T)$, which, of course, is related to $\delta(T)$ through an integral form of the Milne relation, are also listed in Table 2.9. Comparison of Table 2.9 with Table 2.8 shows that charge exchange has a rate comparable with recombination in converting O+ to O⁰ at the typical H II region conditions, while again at the outer edge of the nebula, charge exchange

TABLE 2.9
Charge-Exchange Reaction Coefficients

T (°K)	O		N	
	δ (10^{-9} cm³ sec⁻¹)	δ' (10^{-9} cm³ sec⁻¹)	δ (10^{-9} cm³ sec⁻¹)	δ' (10^{-9} cm³ sec⁻¹)
1000	0.74	1.02	0.0	0.45
3000	0.83	1.03	0.040	0.39
5000	0.87	1.04	0.18	0.38
10,000	0.91	1.04	0.53	0.37

dominates because of the higher density of H^0. At the higher radiation densities that occur in planetary nebulae, charge exchange is not important in the ionization balance of O except near the outer edge of the ionized region. The charge-exchange reactions (2.34) and (2.37) do not appreciably affect the ionization equilibrium of H, because of the low densities of O and O^+.

At temperatures that are high compared with the difference in ionization potentials between O and H, the charge-exchange reactions (2.34) and (2.37) tend to set up an equilibrium in which the ratio of species depends only on the statistical weights, so $\delta/\delta' \to 8/9$. It can be seen from Table 2.9 that this situation is closely realized at $T = 10,000°K$, and thus in a nebula, wherever charge-exchange processes dominate the ionization balance of O, its degree of ionization is locked to that of H by the equation

$$\frac{N_{O^0}}{N_{O^+}} = \frac{9}{8}\frac{N_{H^0}}{N_p}. \tag{2.39}$$

The ionization potential of N (14.5 eV) is also close to that of H (13.60 eV), although it is not so close as that of O (13.61 eV), and charge exchange is also important in its ionization balance in the outer boundaries of ionization-bounded nebulae. The charge-exchange reactions equivalent to (2.34) and (2.37) are

$$N^0(^4S) + H^+ \rightleftarrows N^+(^3P) + H^0(^2S), \tag{2.40}$$

and the calculated reaction rate constants δ (for the reaction proceeding from left to right) and δ' (from right to left) are also listed in Table 2.9. Because of the different statistical weights,

$$\frac{N_{N^0}}{N_{N^+}} \to \frac{\delta'}{\delta}\frac{N_{H^0}}{N_p} \to \frac{2}{9}\frac{N_{H^0}}{N_p} \tag{2.41}$$

at high temperatures, but because of the higher ionization potential of N, the second part of this equation is not closely approached at $T = 10,000°K$.

Though these charge-exchange reactions have not been included in many nebular calculations, in fact there are many situations in which they are important. As will be seen in Section 5.9, the few calculations that do include them agree better with observational data than do the calculations that ignore them.

References

Much of the early work on gaseous nebulae is due to H. Zanstra, D. H. Menzel, L. H. Aller, and others. The very important series of papers on physical processes in gaseous nebulae by Menzel and his collaborators is collected in Menzel, D. H. 1962. *Selected papers on physical processes in ionized nebulae.* New York: Dover. The treatment in this chapter is based on ideas that were, in many cases, given in these pioneering papers. The specific formulation and the numerical values used in this chapter are largely based on the references listed here.

Basic papers on ionization structure:
> Strömgren, B. 1939. *Ap. J.* **89,** 529.
> Hummer, D. G., and Seaton, M. J. 1963. *M. N. R. A. S.* **125,** 437.
> Hummer, D. G., and Seaton, M. J. 1964. *M. N. R. A. S.* **127,** 217.

Numerical values of H recombination coefficient:
> Seaton, M. J. 1959. *M. N. R. A. S.* **119,** 81.
> Burgess, A. 1964. *Mem. R. A. S.* **69,** 1.
> Pengelly, R. M. 1964. *M. N. R. A. S.* **127,** 145.

(Tables 2.1 and 2.6 are based on these references.)

Numerical values of He recombination coefficient:
> Burgess, A., and Seaton, M. J. 1960. *M. N. R. A. S.* **121,** 471.
> Robbins, R. R. 1968. *Ap. J.* **151,** 497.
> Robbins, R. R. 1970. *Ap. J.* **160,** 519.
> Brown, R. L., and Mathews, W. G. 1970. *Ap. J.* **160,** 939.

(Table 2.4 is based on these references.)

Numerical values of He ($2\,^3S \rightarrow 2\,^1S$ and $2\,^1P$) collisional cross sections:
> Burke, P. G., Cooper, J. W., and Ormonde, S. 1969. *Phys. Rev.* **183,** 245.

(Table 2.5 is based on this reference.)

Numerical values of the H and He^+ photoionization cross sections:
> Hummer, D. G., and Seaton, M. J. 1963. *M. N. R. A. S.* **125,** 437.

(Figure 2.2 is based on this reference and the following one.)

Numerical values of the He photoionization cross section:
> Bell, K. L., and Kingston, A. E. 1967. *Proc. Phys. Soc.* **90,** 31.

Numerical values of He ($2\,^1S \rightarrow 1\,^1S$) and He ($2\,^3S \rightarrow 1\,^1S$) transition probabilities:
> Griem, H. R. 1969. *Ap. J.* **156,** L103.
> Drake, G. W. F., Victor, G. A., and Dalgarno, A. 1969. *Phys. Rev.* **180,** 25.
> Drake, G. W. 1971. *Phys. Rev. A* **3,** 908.

Numerical values of heavy-element photoionization cross sections:
     Flower, D. R. 1968. *Planetary Nebulae* (IAU Symposium No. 34), ed. D. E.
          Osterbrock and C. R. O'Dell. Dordrecht: Reidel, p. 205.
     Henry, R. J. W. 1970. *Ap. J.* **161**, 1153.
(Table 2.7 and Figure 2.8 are based on the preceding two references.)

Charge-exchange reaction of O and N:
     Chamberlain, J. W. 1956. *Ap. J.* **124**, 390.
     Field, G. B., and Steigman, G. 1971. *Ap. J.* **166**, 59.
     Steigman, G., Werner, M. W., and Geldon, F. M. 1971. *Ap. J.* **168**, 373.
(Table 2.9 is based on the preceding two references.)

Calculations of model H II regions:
     Hjellming, R. M. 1966. *Ap. J.* **143**, 420.
     Rubin, R. H. 1968. *Ap. J.* **153**, 761.
(Figure 2.4 is based on the latter reference.)

Calculations of model planetary nebulae:
     Harrington, J. P. 1969. *Ap. J.* **156**, 903.
     Flower, D. R. 1969, *M. N. R. A. S.* **146**, 171.
(Figure 2.6 is based on the latter reference.)

Model atmospheres for early-type stars are discussed in more detail in Chapter 5. For the purposes of the present chapter, it is easiest to say that the simplest models are black bodies; a much better approximation is provided by models in which the continuous spectrum is calculated; and the best models are those that also include the effects of the absorption lines, as they are strong and numerous in the ultraviolet.

Models for continuous spectrum:
     Hummer, D. G., and Mihalas, D. 1970. *M. N. R. A. S.* **147**, 339.

Temperature scale and models with line blanketing:
     Morton, D. C. 1969. *Ap. J.* **158**, 629.

3

Thermal Equilibrium

3.1   Introduction

The temperature in a static nebula is fixed by the equilibrium between heating by photoionization and cooling by recombination and by radiation from the nebula. When a photon of energy $h\nu$ is absorbed and causes an ionization of H, the photoelectron produced has an initial energy $(1/2)mv^2 = h(\nu - \nu_0)$, and we may think of an electron being "created" with this energy. The electrons thus produced are rapidly thermalized, as indicated in Chapter 2, and in ionization equilibrium these photoionizations are balanced by an equal number of recombinations. In each recombination, a thermal electron with energy $(1/2)mv^2$ disappears, and an average of this quantity over all recombinations represents the mean energy that "disappears" per recombination. The difference between the mean energy of a newly created photoelectron and the mean energy of a recombining electron represents the net gain in energy by the electron gas per ionization process. In equilibrium this net energy gain is balanced by the energy lost by radiation, chiefly by electron collisional excitation of bound levels of abundant ions, followed by emission of photons that can escape from the nebula. Free-free emission, or bremsstrahlung, is another less important radiative energy-loss mechanism.

In this chapter we shall examine each of these processes, and then the resulting nebular temperatures calculated from them. The processes that are responsible for radiative cooling are, of course, the same processes that are responsible for emission of the observed radiation from the nebula, so these processes will be discussed again from this point of view in the following chapter.

3.2   Energy Input by Photoionization

Let us first examine the energy input by photoionization. As in Chapter 2, it is simplest to begin by considering a pure H nebula. At any particular point in the nebula, the energy input (per unit volume per unit time) is

$$G(\mathrm{H}) = N_{\mathrm{H}^0} \int_{\nu_0}^{\infty} \frac{4\pi J_\nu}{h\nu} h(\nu - \nu_0) a_\nu(\mathrm{H}^0)\, d\nu. \tag{3.1}$$

Furthermore, since the nebula is in ionization equilibrium, we may eliminate N_{H^0} by substituting equation (2.8), giving

$$G(\mathrm{H}) = N_e N_p \alpha_A(\mathrm{H}^0, T) \frac{\displaystyle\int_{\nu_0}^{\infty} \frac{J_\nu}{h\nu} h(\nu - \nu_0) a_\nu(\mathrm{H}^0)\, d\nu}{\displaystyle\int_{\nu_0}^{\infty} \frac{J_\nu}{h\nu} a_\nu(\mathrm{H}^0)\, d\nu}$$

$$= N_e N_p \alpha_A(\mathrm{H}^0, T) \frac{3}{2} kT_i. \tag{3.2}$$

From this equation it can be seen that the mean energy of a newly created photoelectron depends on the form of the ionizing radiation field but not on the absolute strength of the radiation. The rate of creation of photoelectrons depends on the strength of the radiation field, or, as equation (3.2) shows, on the recombination rate. The quantity $(3/2)kT_i$ represents the initial temperature of the newly created photoelectrons. For assumed black-body spectra with $J_\nu = B_\nu(T_*)$, it is easy to show that $T_i \approx T_*$ so long as $kT_* < h\nu_0$. For any known J_ν (for instance, the emergent spectrum from a model atmosphere), the integration can be carried out numerically, and a short list of representative values of T_i is given in Table 3.1. Note that column two in the table, $\tau_0 = 0$, corresponds to photoionization by the emergent model-atmosphere spectrum. At larger distances from the star, the spectrum of the ionizing radiation is modified by absorption in the nebula, the radiation nearest the series limit being most strongly attenuated because of the

TABLE 3.1
Mean Input Energy of Photoelectrons

Model stellar atmosphere, T_*	T_i			
	$\tau_0 = 0$	$\tau_0 = 1$	$\tau_0 = 5$	$\tau_0 = 10$
3.0×10^4	1.46×10^4	1.81×10^4	3.61×10^4	5.45×10^4
3.5×10^4	2.15×10^4	2.65×10^4	4.67×10^4	6.31×10^4
4.0×10^4	2.67×10^4	3.38×10^4	6.52×10^4	9.57×10^4
5.0×10^4	3.50×10^4	4.47×10^4	8.47×10^4	11.87×10^4

frequency dependence of the absorption coefficient. Therefore, the higher-energy photons penetrate further into the gas, and the mean energy of the photoelectrons produced at larger optical depths from the star is higher. This effect is shown for a pure H nebula in the columns labeled with values of τ_0, the optical depth at the ionization limit.

3.3 Energy Loss by Recombination

The kinetic energy lost by the electron gas (per unit volume per unit time) in recombination can be written

$$L_R(H) = N_e N_p kT \beta_A(H^0, T),$$ (3.3)

where

$$\beta_A(H^0, T) = \sum_{n=1}^{\infty} \beta_n(H^0, T) = \sum_{n=1}^{\infty} \sum_{L=0}^{n-1} \beta_{nL}(H^0, T),$$ (3.4)

with

$$\beta_{nL}(H^0, T) = \frac{1}{kT} \int_0^{\infty} v\sigma_{nL}(H^0, T) \frac{1}{2} mv^2 f(v)\, dv;$$ (3.5)

the left-hand side of equation (3.5) is thus effectively a kinetic energy-averaged recombination coefficient. Note that since the recombination cross sections are approximately proportional to v^{-2}, the electrons of lower kinetic energy are preferentially captured, and the mean energy of the captured electrons is somewhat less than $3kT/2$. Calculated values of β_1 and β_A are listed in Table 3.2.

In a pure H nebula that had no radiation losses, the thermal equilibrium

Recombination Cooling Coefficient

$T\,(^\circ\mathrm{K})$	β_A (cm³ sec⁻¹)	β_1 (cm³ sec⁻¹)	β_B (cm³ sec⁻¹)
2500	8.93×10^{-13}	3.13×10^{-13}	5.80×10^{-13}
5000	5.42×10^{-13}	2.20×10^{-13}	3.22×10^{-13}
10,000	3.23×10^{-13}	1.50×10^{-13}	1.73×10^{-13}
20,000	1.88×10^{-13}	9.58×10^{-14}	9.17×10^{-14}

equation would be

$$G(\mathrm{H}) = L_R(\mathrm{H}), \tag{3.6}$$

and the solution for the nebular temperature would give a $T > T_i$ because of the "heating" due to the capture of the slower electrons.

The radiation field J_ν in equation (3.1) should, of course, include the diffuse radiation as well as the stellar radiation modified by absorption. This can easily be included in the on-the-spot approximation, since, according to it, every emission of an ionizing photon in recombination to the level $n = 1$ is balanced by absorption of the same photon at a nearby spot in the nebula. Thus production of photons by the diffuse radiation field and recombinations to the ground level can simply be omitted from the gain and loss rates, leading to the equations

$$G_{OTS}(\mathrm{H}) = N_{\mathrm{H}^0} \int_{\nu_0}^\infty \frac{4\pi J_{\nu s}}{h\nu} h(\nu - \nu_0) a_\nu(\mathrm{H}^0)\, d\nu$$

$$= N_e N_p \alpha_B(\mathrm{H}^0, T)\, \frac{\int_{\nu_0}^\infty \dfrac{J_{\nu s}}{h\nu} h(\nu - \nu_0) a_\nu(\mathrm{H}^0)\, d\nu}{\int_{\nu_0}^\infty \dfrac{J_{\nu s}}{h\nu} a_\nu(\mathrm{H}^0)\, d\nu} \tag{3.7}$$

and

$$L_{OTS}(\mathrm{H}) = N_e N_p kT \beta_B(\mathrm{H}^0, T), \tag{3.8}$$

with

$$\beta_B(\mathrm{H}^0, T) = \sum_{n=2}^\infty \beta_n(\mathrm{H}^0, T). \tag{3.9}$$

The generalization to include He in the heating and recombination cooling rates is straightforward to write, namely,

$$G = G(\text{H}) + G(\text{He}), \tag{3.10}$$

where

$$G(\text{He}) = N_e N_{\text{He}^+} \alpha_A(\text{He}^0, T) \frac{\displaystyle\int_{\nu_2}^{\infty} \frac{J_\nu}{h\nu} h(\nu - \nu_2) a_\nu(\text{He}^0)\, d\nu}{\displaystyle\int_{\nu_2}^{\infty} \frac{J_\nu}{h\nu} a_\nu(\text{He}^0)\, d\nu} \tag{3.11}$$

and

$$L_R = L_R(\text{H}) + L_R(\text{He}), \tag{3.12}$$

with

$$L_R(\text{He}) = N_e N_{\text{He}^+} kT \beta_A(\text{He}^0, T), \tag{3.13}$$

and so on.

It can be seen that the heating and recombination cooling rates are proportional to the densities of the ions involved, and therefore the contributions of the heavy elements, which are much less abundant than H and He, can, to a good approximation, be omitted from these rates.

3.4   Energy Loss by Free-Free Radiation

Next we shall examine cooling by radiation not involving recombination, which, in most circumstances, is far larger than the recombination cooling and therefore dominates the thermal equilibrium. A minor contributor to the cooling rate, which is important only because it can occur in a pure H nebula, is free-free radiation or bremsstrahlung, in which a continuous spectrum is emitted. The rate of cooling by this process by ions of charge Z, integrated over all frequencies, is, to a fair approximation,

$$
\begin{aligned}
L_{FF}(Z) &= 4\pi j_{ff} \\
&= \frac{2^5 \pi e^6 Z^2}{3^{3/2} hmc^3} \left(\frac{2\pi kT}{m}\right)^{1/2} N_e N_+ \\
&= 1.42 \times 10^{-27}\, Z^2\, T^{1/2} N_e N_+ \tag{3.14}
\end{aligned}
$$

in ergs cm^{-3} sec^{-1}. Again H$^+$ dominates the free-free cooling, because of its abundance, and He$^+$ can be included with H$^+$ (since both have $Z = 1$) by writing $N_+ = N_p + N_{\text{He}^+}$.

3.5 Energy Loss by Collisionally Excited Line Radiation

A far more important source of radiative cooling is collisional excitation of low-lying energy levels of common ions, such as O^+, O^{++}, and N^+. These ions make a significant contribution in spite of their low abundance because they have energy levels with excitation potentials of the order of kT, while H and He only have levels with considerably higher excitation potential, and are therefore not important collisionally excited coolants under most circumstances. Let us therefore examine the excitation to level 2 of a particular ion by electron collisions with ions in the lower level 1. The cross section for excitation $\sigma_{12}(v)$ is a function of electron velocity and is zero below the threshold $\chi = h\nu_{21}$. Not too far above the threshold, the main dependence of the excitation cross section is $\sigma \propto v^{-2}$ (due to the focusing effect of the Coulomb force), and it is therefore convenient to express the collision cross sections in terms of the collision strength $\Omega(1, 2)$ defined by

$$\sigma_{12}(v) = \frac{\pi\hbar^2}{m^2v^2} \frac{\Omega(1, 2)}{\omega_1} \quad \text{for } \frac{1}{2}mv^2 > \chi, \tag{3.15}$$

where $\Omega(1, 2)$ is a function of electron velocity (or energy) but is often approximately constant near the threshold, and ω_1 is the statistical weight of the lower level.

There is a relation between the cross section for de-excitation, $\sigma_{21}(v)$, and the cross section for excitation, namely,

$$\omega_1 v_1^2 \sigma_{12}(v_1) = \omega_2 v_2^2 \sigma_{21}(v_2), \tag{3.16}$$

where v_1 and v_2 are related by

$$\frac{1}{2}mv_1^2 = \frac{1}{2}mv_2^2 + \chi. \tag{3.17}$$

Equation (3.16) can easily be derived from the principle of detailed balancing, which states that in thermodynamic equilibrium each microscopic process is balanced by its inverse. Thus in this particular case, the number of excitations caused by collisions with electrons in the velocity range v_1 to $v_1 + dv_1$ is just balanced by the de-excitations caused by collisions that produce electrons in the same velocity range. Thus

$$N_e N_1 v_1 \sigma_{12}(v_1) f(v_1) \, dv_1 = N_e N_2 v_2 \sigma_{21}(v_2) f(v_2) \, dv_2,$$

and using the Boltzmann equation of thermodynamic equilibrium,

$$\frac{N_2}{N_1} = \frac{\omega_2}{\omega_1} e^{-\chi/kT},$$

we derive the relation (3.16). Combining equations (3.15) and (3.16) so that the de-excitation cross section can be expressed in terms of the collision strength $\Omega(1, 2)$,

$$\sigma_{21}(v_2) = \frac{\pi\hbar^2}{mv_2^2}\frac{\Omega(1, 2)}{\omega_2};$$
(3.18)

that is, the collision strengths are symmetrical in 1 and 2.

The total collisional de-excitation rate per unit volume per unit time is thus

$$N_e N_2 q_{21} = N_e N_2 \int_0^\infty v\,\sigma_{21}(v) f(v)\, dv$$

$$= N_e N_2 \left(\frac{2\pi}{kT}\right)^{1/2} \frac{\hbar^2}{m^{3/2}} \frac{\Omega(1, 2)}{\omega_2}$$

$$= N_e N_2 \frac{8.629 \times 10^{-6}}{T^{1/2}} \frac{\Omega(1, 2)}{\omega_2}$$
(3.19)

(in $cm^3\ sec^{-1}$) if $\Omega(1, 2)$ is a constant. In general, the mean value

$$\Omega(1, 2) = \int_0^\infty \Omega(1, 2; E)\, e^{-E/kT} d\left(\frac{E}{kT}\right)$$
(3.20)

should be used in equation (3.19), where $E = (1/2)mv_2^2$. Likewise, the collisional excitation rate per unit volume per unit time is $N_e N_1 q_{12}$, where

$$q_{12} = \frac{\omega_2}{\omega_1} q_{21}\, e^{-\chi/kT}.$$
(3.21)

The collision strengths must be calculated quantum mechanically, and selected lists of some of the most important numerical values are given in Tables 3.3 through 3.7. A particular collision strength in general consists of a part that varies slowly with energy, on which, in some cases, there are superimposed contributions that vary rapidly with energy; but when the cross sections are integrated over a Maxwellian distribution, as in almost all astrophysical applications, the effect of the exact positions of the resonances tends to be averaged out. Therefore, it is sufficient in most cases to use the collision strength averaged over resonances, and as this is considerably simpler to calculate and to list, it is the quantity given in Tables 3.3–3.7, evaluated at $T = 10,000°K$, a representative nebular temperature. It is convenient to remember that, for an electron with the mean energy

TABLE 3.3
Collision Strengths for p and p⁵ Ions

Ion	$\Omega(^2P_{1/2}, {}^2P_{3/2})$	Ion	$\Omega(^2P_{1/2}, {}^2P_{3/2})$
C^+	1.33	Si^+	7.7
N^{+2}	1.09	P^{+2}	1.84
O^{+3}	0.76	S^{+3}	1.30
Ne^{+5}	0.37	Ar^{+5}	0.70
Ne^+	0.37	Ar^+	0.78
Mg^{+3}	0.31	Ca^{+3}	0.82
Si^{+5}	0.23		

TABLE 3.4
Collision Strengths for p² and p⁴ Ions

Ion	$\Omega(^3P, {}^1D)$	$\Omega(^3P, {}^1S)$	$\Omega(^1D, {}^1S)$	$\Omega(^3P_0, {}^3P_1)$	$\Omega(^3P_0, {}^3P_2)$	$\Omega(^3P_1, {}^3P_2)$
N^+	2.99	0.36	0.39	0.41	0.28	1.38
O^{+2}	2.50	0.30	0.58	0.39	0.21	0.95
Ne^{+4}	1.47	0.20	0.19	0.23	0.11	0.53
Ne^{+2}	1.14	0.17	0.20	0.20	0.13	0.54
S^{+2}	3.87	0.75	1.34	0.94	0.51	2.32
Ar^{+4}	0.82	0.10	1.02	0.20	0.29	0.90
Ar^{+2}	3.92	0.59	0.84	0.86	0.42	2.02

TABLE 3.5
Collision Strengths for p³ Ions

Ion	$\Omega(^4S, {}^2D)$	$\Omega(^4S, {}^2P)$	$\Omega(^2D_{3/2}, {}^2D_{5/2})$	$\Omega(^2D_{3/2}, {}^2P_{1/2})$
O^+	1.47	0.45	1.16	0.29
Ne^{+3}	1.14	0.43	0.80	0.22
S^+	5.66	2.72	5.55	1.96
Ar^{+3}	1.67	0.65	2.31	0.86

Ion	$\Omega(^2D_{3/2}, {}^2P_{3/2})$	$\Omega(^2D_{5/2}, {}^2P_{1/2})$	$\Omega(^2D_{5/2}, {}^2P_{3/2})$	$\Omega(^2P_{1/2}, {}^2P_{3/2})$
O^+	0.44	0.32	0.78	0.28
Ne^{+3}	0.36	0.26	0.61	0.35
S^+	2.82	2.02	5.15	1.33
Ar^{+3}	1.10	0.74	2.17	0.62

TABLE 3.6
Collision Strengths for ^2S–^2P Transitions

Ion	$\Omega(2s\,^2S,\,2p\,^2P)$	Ion	$\Omega(3s\,^2S,\,3p\,^2P)$
C^{+3}	9.76	Mg^+	20.2
N^{+4}	7.08	Si^{+3}	19.1
O^{+5}	5.28		

TABLE 3.7
Collision Strengths for ^1S–^3P Transitions

Ion	$\Omega(2s^2\,^1S,\,2s2p\,^3P)$	Ion	$\Omega(3s^2\,^1S,\,3s3p\,^3P)$
C^{+2}	1.41	Si^{+2}	2.
N^{+3}	0.75		
O^{+4}	0.49		
Ne^{+6}	0.26		

at a typical nebular temperature, $T \approx 7500°K$, the cross sections for excitation and de-excitation are $\sigma \approx 10^{-15}\Omega/\omega$ cm^2.

It should be noted that there is a simple relation among the collision strengths between a term consisting of single level and a term consisting of various levels, namely,

$$\Omega(SLJ,\,S'L'J') = \frac{(2J' + 1)}{(2S' + 1)(2L' + 1)}\,\Omega(SL,\,S'L') \qquad (3.22)$$

if either $S = 0$ or $L = 0$. The factors $(2J' + 1)$ and $(2S' + 1)(2L' + 1)$ will be recognized as the statistical weights of the level and of the term, respectively. On account of this relation, the rate of collisional excitation in p^2 or p^4 ions (such as O^{++}) from the ground ^3P term to the excited (singlet) ^1D and ^1S levels is very nearly independent of the distribution of ions among $^3P_0$, $^3P_1$, and $^3P_2$.

For all the low-lying levels of the ions listed in Tables 3.3 through 3.5, the excited levels arise from the same electron configuration as the ground level. Radiative transitions between excited levels and the ground level are therefore forbidden by the electric-dipole selection rules, but can occur by magnetic dipole and/or electric quadrupole transitions. These are the well-known forbidden lines, many of which are observed in nebular spectra, while others are too far in the infrared or ultraviolet to be observable from the ground. Transition probabilities, as well as wavelengths for the observable lines, are listed in Tables 3.8 through 3.10.

In the simple case of an ion with a single excited level, in the limit of very low electron density every collisional excitation is followed by the emission of a photon, and the cooling rate per unit volume is therefore

$$L_C = N_e N_1 q_{12} h\nu_{21}. \tag{3.23}$$

However, if the density is sufficiently high, collisional de-excitation is not negligible and the cooling rate is reduced. The equilibrium equation for the balance between the excitation and de-excitation rates of the excited level is, in general,

$$N_e N_1 q_{12} = N_e N_2 q_{21} + N_2 A_{21}, \tag{3.24}$$

and the solution is

$$\frac{N_2}{N_1} = \frac{N_e q_{12}}{A_{21}} \left[\frac{1}{1 + \dfrac{N_e q_{21}}{A_{21}}} \right], \tag{3.25}$$

so the cooling rate is

$$L_C = N_2 A_{21} h\nu_{21} = N_e N_1 q_{12} h\nu_{21} \left[\frac{1}{1 + \dfrac{N_e q_{21}}{A_{21}}} \right]. \tag{3.26}$$

It can be seen that as $N_e \to 0$, we recover equation (3.23), while as $N_e \to \infty$,

$$L_C \to N_1 \frac{\omega_2}{\omega_1} e^{-\chi/kT} A_{21} h\nu_{21}, \tag{3.27}$$

the thermodynamic-equilibrium cooling rate.

Some ions have only two low-lying levels and may be treated by this simple formalism, but most ions have more levels, and in particular all ions with ground configurations p^2, p^3, or p^4 have five low-lying levels. Examples are O^{++} and N^+, with energy-level diagrams shown in Figure 3.1. In such cases, collisional and radiative transitions can occur between any of the levels, and excitation and de-excitation cross sections and collision strengths exist between all pairs of the levels.

TABLE 3.8
Transition Probabilities of p^2 Ions

Transition	[N II] Transition probability (sec⁻¹)	Wave-length (Å)	[O III] Transition probability (sec⁻¹)	Wave-length (Å)	[Ne V] Transition probability (sec⁻¹)	Wave-length (Å)	[S III] Transition probability (sec⁻¹)	Wave-length (Å)	[Ar V] Transition probability (sec⁻¹)	Wave-length (Å)
$^1D_2-^1S_0$	1.1	5754.6	1.6	4363.2	2.6	2974.8	2.5	6312.1	3.8	4625.5
$^3P_2-^1S_0$	1.6×10^{-4}	3070.8	7.1×10^{-4}	2331.4	6.8×10^{-3}	1592.7	1.6×10^{-2}	3797.4	8.1×10^{-2}	2786.0
$^3P_1-^1S_0$	3.4×10^{-2}	3062.8	2.3×10^{-1}	2321.0	4.2	1575.2	8.5×10^{-1}	3721.7	6.8	2691.1
$^3P_2-^1D_2$	3.0×10^{-3}	6583.4	2.1×10^{-2}	5006.9	3.8×10^{-1}	3425.9	6.4×10^{-2}	9531.8	5.1×10^{-1}	7005.7
$^3P_1-^1D_2$	1.0×10^{-3}	6548.1	7.1×10^{-3}	4958.9	1.4×10^{-1}	3345.8	2.5×10^{-2}	9069.0	2.2×10^{-1}	6435.1
$^3P_0-^1D_2$	4.2×10^{-7}	6527.1	1.9×10^{-6}	4931.0	1.9×10^{-5}	3300.1	9.1×10^{-6}	8830.9	4.9×10^{-5}	6133.1
$^3P_1-^3P_2$	7.5×10^{-6}	—	9.8×10^{-5}	—	4.6×10^{-3}	—	2.4×10^{-3}	—	2.7×10^{-2}	—
$^3P_0-^3P_2$	1.3×10^{-12}	—	3.5×10^{-11}	—	5.2×10^{-9}	—	4.7×10^{-8}	—	1.3×10^{-6}	—
$^3P_0-^3P_1$	2.1×10^{-6}	—	2.6×10^{-5}	—	1.3×10^{-3}	—	4.7×10^{-4}	—	8.0×10^{-3}	—

TABLE 3.9
Transition Probabilities of p^3 Ions

Transition	[O II] Transition probability (sec⁻¹)	[O II] Wavelength (Å)	[Ne IV] Transition probability (sec⁻¹)	[Ne IV] Wavelength (Å)	[S II] Transition probability (sec⁻¹)	[S II] Wavelength (Å)	[Ar IV] Transition probability (sec⁻¹)	[Ar IV] Wavelength (Å)
$^2P_{1/2}-^2P_{3/2}$	6.0×10^{-11}	—	2.3×10^{-9}	—	1.0×10^{-6}	—	5.2×10^{-5}	—
$^2D_{5/2}-^2P_{3/2}$	1.2×10^{-1}	7319.9	4.0×10^{-1}	4714.3	2.1×10^{-1}	10320.5	6.7×10^{-1}	7237.3
$^2D_{3/2}-^2P_{3/2}$	6.1×10^{-2}	7330.2	4.4×10^{-1}	4724.2	1.7×10^{-1}	10286.7	9.1×10^{-1}	7170.6
$^2D_{5/2}-^2P_{1/2}$	6.1×10^{-2}	7319.9	1.1×10^{-1}	4715.6	8.7×10^{-2}	10370.5	1.2×10^{-1}	7331.4
$^2D_{3/2}-^2P_{1/2}$	1.0×10^{-1}	7330.2	3.9×10^{-1}	4725.6	2.0×10^{-1}	10336.4	6.8×10^{-1}	7262.8
$^4S_{3/2}-^2P_{3/2}$	6.0×10^{-2}	2470.3	1.3	1602.0	3.4×10^{-1}	4068.6	2.6	2853.6
$^4S_{3/2}-^2P_{1/2}$	2.4×10^{-2}	2470.3	5.3×10^{-1}	1602.1	1.3×10^{-1}	4076.4	9.7×10^{-1}	2868.2
$^2D_{5/2}-^2D_{3/2}$	1.3×10^{-7}	—	1.4×10^{-6}	—	3.3×10^{-7}	—	2.3×10^{-5}	—
$^4S_{3/2}-^2D_{5/2}$	4.2×10^{-5}	3728.8	5.9×10^{-4}	2425.4	4.7×10^{-4}	6716.4	2.2×10^{-3}	4711.3
$^4S_{3/2}-^2D_{3/2}$	1.8×10^{-4}	3726.1	5.6×10^{-3}	2422.8	1.8×10^{-3}	6730.8	2.8×10^{-2}	4740.2

TABLE 3.10
Transition Probabilities of p^4 Ions

Transition	[O I]		[Ne III]		[Ar III]	
	Transition probability (sec^{-1})	Wavelength (A)	Transition probability (sec^{-1})	Wavelength (A)	Transition probability (sec^{-1})	Wavelength (A)
$^1D_2-^1S_0$	1.3	5577.4	2.8	3342.5	3.1	5191.8
$^3P_2-^1S_0$	3.7×10^{-4}	2958.4	5.1×10^{-3}	1793.7	4.3×10^{-2}	3005.2
$^3P_1-^1S_0$	6.7×10^{-2}	2972.3	2.2	1814.7	4.0	3109.2
$^3P_2-^1D_2$	5.1×10^{-3}	6300.3	1.7×10^{-1}	3868.8	3.2×10^{-1}	7135.8
$^3P_1-^1D_2$	1.64×10^{-3}	6363.8	5.2×10^{-2}	3967.5	8.3×10^{-2}	7751.1
$^3P_0-^1D_2$	1.1×10^{-6}	6391.5	1.2×10^{-5}	4011.6	2.9×10^{-5}	8036.3
$^3P_1-^3P_0$	1.7×10^{-5}	—	1.2×10^{-3}	—	5.1×10^{-3}	—
$^3P_2-^3P_0$	1.0×10^{-10}	—	2.0×10^{-8}	—	2.7×10^{-6}	—
$^3P_2-^3P_1$	9.0×10^{-5}	—	6.0×10^{-3}	—	3.1×10^{-2}	—

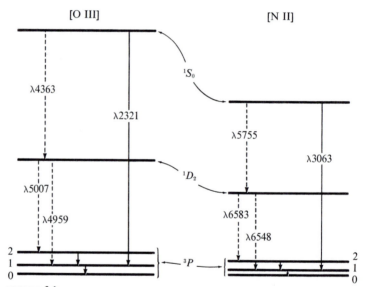

FIGURE 3.1
Energy-level diagram for lowest terms of [O III], all from ground $2p^2$
configuration, and for [N II], of the same isoelectronic sequence. Splitting of
the ground ^3P term has been exaggerated for clarity. Emission lines in the
optical region are indicated by dashed lines and by solid lines in the infrared
and ultraviolet. Only the strongest transitions are indicated.

The equilibrium equations for each of the levels $i = 1, 5$ thus become

$$\sum_{j \neq i} N_j N_e q_{ji} + \sum_{j > i} N_j A_{ji} = \sum_{j \neq i} N_i N_e q_{ij} + \sum_{j < i} N_i A_{ij}, \qquad (3.28)$$

which, together with the total number of ions

$$\sum_j N_j = N, \qquad (3.29)$$

can be solved for the relative population in each level, and then for the collisionally excited radiative cooling rate

$$L_C = \sum_i N_i \sum_{j < i} A_{ij} h\nu_{ij}. \qquad (3.30)$$

In the low-density limit, $N_e \to 0$, this becomes a sum of terms like (3.23), but if

$$N_e q_{ij} > \sum_{k < i} A_{ik}$$

for any i, j, collisional de-excitation is not negligible and the complete solution must be used. In fact, for any level i, a critical density $N_c(i)$ may be defined as

$$N_c(i) = \sum_{j < i} A_{ij} \Big/ \sum_{j \neq i} q_{ij}, \qquad (3.31)$$

so that for $N_e < N_c(i)$, collisional de-excitation of level i is negligible, while for $N_e > N_c(i)$, it is important. A list of critical densities for levels that are most important in radiative cooling is given in Table 3.11.

TABLE 3.11
Critical Densities for Collisional De-excitation

Ion	Level	N_c (cm^{-3})	Ion	Level	N_c (cm^{-3})
C II	$^2P_{3/2}$	7.6×10	O III	$^1D_2$	6.5×10^5
			O III	$^3P_2$	4.9×10^3
C III	$^3P_2$	2.6×10^5	O III	$^3P_1$	6.7×10^2
N II	$^1D_2$	7.8×10^4	Ne II	$^2P_{1/2}$	7.7×10^5
N II	$^3P_2$	2.6×10^2			
N II	$^3P_1$	4.1×10	Ne III	$^1D_2$	1.1×10^6
			Ne III	$^3P_0$	4.2×10^4
N III	$^2P_{3/2}$	2.0×10^3	Ne III	$^3P_1$	2.8×10^5
N IV	$^3P_2$	9.1×10^5	Ne V	$^1D_2$	1.1×10^6
			Ne V	$^3P_2$	4.2×10^5
O II	$^2D_{3/2}$	3.3×10^3	Ne V	$^3P_1$	1.0×10^5
O II	$^2D_{5/2}$	6.3×10^2			

NOTE: All values are calculated for $T = 10,000°$K.

3.6   Energy Loss by Collisionally Excited Line Radiation of H

H^+, the most abundant ion in nebulae, of course has no bound levels and no lines, but H^0, although its fractional abundance is low, may affect the radiative cooling in a nebula. The most important excitation processes from the ground $1\,^2S$ term are to $2\,^2P$, followed by emission of an $L\alpha$ photon with $h\nu = 10.2$ eV, and to $2\,^2S$, followed by emission of two photons in the $2\,^2S \rightarrow 1\,^2S$ continuum with $h\nu' + h\nu'' = 10.2$ eV and transition probability $A_{2\,^2S,\,1\,^2S} = 8.23$ sec^{-1}. Cross sections for excitation of neutral atoms by electrons do not vary as v^{-2}, but rise from zero at the threshold, peak at energies several times the threshold, and then decline at high energies, often with superimposed resonances; so it is not particularly advantageous to define a collision strength for them, but instead the computed cross sections can be integrated over a Maxwellian distribution directly to find the excitation rate

$$q_{1,j} = \int_{(1/2)mv^2 = \chi}^{\infty} v\sigma_{1j}(v)f(v)\,dv. \tag{3.32}$$

Numerical values of $q_{1,j}(T)$ are presented in Table 3.12 for $j = 2\,^2S$, $2\,^2P$, $3\,^2S$, $3\,^2P$, and $3\,^2D$, using the most accurate available calculations of cross sections. Because of the large threshold, these excitation rates are not appreciable at low temperatures, but may become important at large T. Accurate cross sections are not available for $n > 3$, and therefore only rough estimates of the radiative cooling by higher levels of H are available, but they seem to make only a minor contribution except at very high temperatures.

TABLE 3.12
Collisional Excitation Rates* for H I

T	$q_{1\,^2S,\,2\,^2S}$	$q_{1\,^2S,\,2\,^2P}$	$q_{1\,^2S,\,3\,^2S}$	$q_{1\,^2S,\,3\,^2P}$	$q_{1\,^2S,\,3\,^2D}$
6×10^3	4.2×10^{-17}	7.0×10^{-17}	3.8×10^{-19}	6.6×10^{-19}	3.5×10^{-19}
8×10^3	5.3×10^{-15}	9.0×10^{-15}	1.3×10^{-16}	2.4×10^{-16}	1.2×10^{-16}
10×10^3	9.4×10^{-14}	1.6×10^{-13}	4.4×10^{-15}	8.1×10^{-15}	3.9×10^{-15}
12×10^3	6.2×10^{-13}	1.1×10^{-12}	4.5×10^{-14}	8.6×10^{-14}	3.9×10^{-14}
14×10^3	2.4×10^{-12}	4.4×10^{-12}	2.3×10^{-13}	4.7×10^{-13}	2.0×10^{-13}
16×10^3	6.4×10^{-12}	1.2×10^{-11}	7.9×10^{-13}	1.7×10^{-12}	6.8×10^{-13}
18×10^3	1.4×10^{-11}	2.7×10^{-11}	2.0×10^{-12}	4.5×10^{-12}	1.7×10^{-12}
20×10^3	2.5×10^{-11}	5.0×10^{-11}	4.3×10^{-12}	9.8×10^{-12}	3.7×10^{-12}

* In cm^3 sec^{-1}.

3.7   Resulting Thermal Equilibrium

The temperature at each point in a static nebula is determined by the equilibrium between heating and cooling rates, namely,

$$G = L_R + L_{FF} + L_C. \tag{3.33}$$

The collisional excited radiative cooling rate L_C is a sum (over all transitions of all ions) of individual terms like (3.23), (3.26), or (3.30), and it can be seen that, in the low-density limit, since all the terms in G, L_R, L_{FF} and L_C are proportional to N_e and to the density of some ion, the equation and therefore the resulting temperature are independent of the total density, but do depend on the relative abundances of the various ions—that is, on the relative abundances of the various elements and on their states of ionization. At higher densities, when collisional de-excitation begins to be important, the cooling rate at a given temperature is decreased, and the equilibrium temperature for a given radiation field is therefore somewhat increased.

To gain a better understanding of the ideas involved, we shall examine a specific case, namely, an H II region with "normal" abundances of the elements. We shall adopt $N(O)/N(H) = 6 \times 10^{-4}$, $N(Ne)/N(H) = 6 \times 10^{-5}$, and $N(N)/N(H) = 5 \times 10^{-5}$. Let us suppose that O, Ne, and N are each 80 percent singly ionized and 20 percent doubly ionized, that H is 0.1 percent neutral, and that the remainder is ionized. Some of the individual contributions to the radiative cooling (in the low-density limit) and the total radiative cooling $L_C + L_{FF}$ are shown in Figure 3.2. For each level the contribution is small if $kT \ll \chi$, then increases rapidly and peaks at $kT \approx \chi$, and then decreases slowly for $kT > \chi$. The total radiative cooling, composed of the sum of the individual contributions, continues to rise with increasing T as long as there are levels with excitation energy $\chi > kT$. It can be seen that, for the assumed composition and ionization, O^{++} dominates the radiative cooling contribution at low temperatures, and O^+ at somewhat higher temperatures. At all temperatures shown, the contribution of collisional excitation of H^0 is small.

It is convenient to rewrite equation (3.33) in the form

$$G - L_R = L_{FF} + L_C,$$

where $G - L_R$ is then the "effective heating rate," representing the net energy gained in photoionization processes, with the recombination losses already subtracted. This effective heating rate is also shown in Figure 3.2, for model stellar atmospheres with a range of temperatures, and it can be seen that the calculated nebular temperature at which the curves cross and equation (3.33) is satisfied is rather insensitive to the input stellar radiation field. Typical nebular temperatures are $T \approx 7000°K$, according to Figure 3.2, with somewhat higher temperatures for hotter stars or larger optical depths.

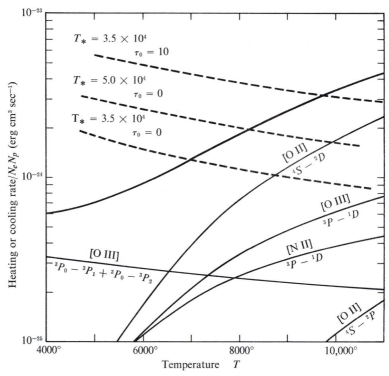

FIGURE 3.2

Net effective heating rates $(G - L_R)$ for various stellar input spectra, shown as dashed curves. Total radiative cooling rate $(L_{FF} + L_C)$ for the simple approximation to the H II region described in the text is shown as heavy solid black curve, and the most important individual contributors to radiative cooling are shown by lighter solid curves. The equilibrium temperature is given by the intersection of a dashed curve and the heavy solid curve. Note how the increased optical depth τ_0 or increased stellar temperature T_* increases T by increasing G.

At high electron densities, collisional de-excitation can appreciably modify the radiative cooling rate and therefore the resulting nebular temperature. For instance, at $N_e \approx 10^4 \, \text{cm}^{-3}$, a density that occurs in condensations in many H II regions, the [O II] $^4S-^2D$ and [O III] $^3P_0-^3P_1$ and $^3P_0-^3P_2$ transitions are only approximately 10 percent effective, while [N II] $^3P_0-^3P_1$ and $^3P_0-^3P_2$ are only about 1 percent effective, and [N III] $^2P_{1/2}-^2P_{3/2}$ are approximately 20 percent effective, as Table 3.11 shows. Figure 3.3 shows the effective cooling rate for this situation, with the abundances and ionization otherwise as previously described, and demonstrates that appreciably higher temperatures occur at high densities.

Under conditions of very high ionization, however, as in the central part of a planetary nebula, the ionization is high enough so that there is very

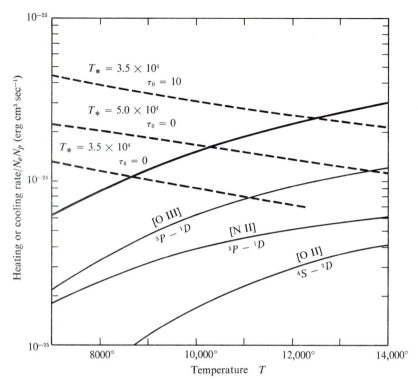

FIGURE 3.3

Same description as for Figure 3.2, except that collisional de-excitation at $N_e = 10^4$ cm^{-3} has been approximately taken into account in the radiative cooling rates.

little H^0, O^+, or O^{++}, and then the radiative cooling is appreciably decreased. Under these conditions the main coolants are Ne^{+4} and C^{+3}, and the nebular temperature may be $T \lesssim 2 \times 10^4$. Detailed results obtained from models of both H II regions and planetary nebulae are discussed in Chapter 5.

References

The basic papers on thermal equilibrium are:
Spitzer, L. 1948. *Ap. J.* **107**, 6.
Spitzer, L. 1949. *Ap. J.* **109**, 337.
Spitzer, L., and Savedoff, M. P. 1950. *Ap. J.* **111**, 593.

Some additional work, including the on-the-spot approximation and the effects of collisional de-excitation, is described in

Burbidge, G. R., Gould, R. J., and Pottasch, S. R. 1963. *Ap. J.* **138**, 945.

Osterbrock, D. E. 1965. *Ap. J.* **142**, 1423.

Numerical values of recombination coefficients β are given in

Hummer, D. G., and Seaton, M. J. 1963. *M. N. R. A. S.* **125**, 437.

(Table 3.2 is based on this reference.)

Basic papers on collisional excitation and the methods used to calculate the collision strengths are:

Seaton, M. J. 1958. *Rev. Mod. Phys.* **30**, 979.

Seaton, M. J. 1968. *Advances in Atomic and Molecular Physics* **4**, 331.

Numerical values of collision strengths are widely scattered through the physics literature, but probably the most accurate values published at the time of writing are:

Saraph, H. E., Seaton, M. J., and Shemming, J. 1969. *Phil. Trans. Roy. Soc. Lon. A* **264**, 77 (all $2p^n$ ions).

Krueger, T. K., and Czyzak, S. J. 1970. *Proc. Roy. Soc. Lon. A* **318**, 531 ($3p^n$ except $3p^3$ is not given).

Czyzak, S. J., Krueger, T. K., Martins, P. de A. P., Saraph, H. E., and Seaton, M. J. 1970. *M. N. R. A. S.* **148**, 361 ($3p^3$ ions).

Saraph, H. E., and Seaton, M. J. 1970. *M. N. R. A. S.* **148**, 367 (S^+ and Cl^{++}).

Burke, P., and Moores, D. 1968. *J. Phys. B* **1**, 575 (Mg^+).

Bely, O. 1966. *Proc. Phys. Soc.* **88**, 587 ($2s$–$2p$ ions).

Bely, O., Tully, J., and van Regemorter, H. 1963. *Ann. Physique* **8**, 303 (Si^{+3}).

Osterbrock, D. E. 1970. *J. Phys. B* (*Proc. Phys. Soc. 1*) **3**, 149 ($2s^2$–$2s2p$ ions).

Eissner, W., Martins, P. de A. P., Nussbaumer, H., Saraph, H. E., and Seaton, M. J. 1969. *M. N. R. A. S.* **146**, 63 (O^+ and O^{++}).

Saraph, H. E., and Seaton, M. J. 1971. *Phil. Trans. Roy. Soc. Lon. A* **271**, 1 (C^+ and Ne^+).

Flower, D. R., and Launay, J. M. 1972. *J. Phys. B.* (*Atom. Molec. Phys.*) **5**, L207 (C^{+2}).

(Tables 3.3–3.7 are based on the preceding eleven references.)

Numerical values of transition probabilities are very conveniently listed in

Garstang, R. H. 1968. *Planetary Nebulae* (IAU Symposium No. 34), ed. D. E. Osterbrock and C. R. O'Dell. Dordrecht: Reidel, p. 143.

This reference, on which Tables 3.8–3.10 are based, also lists the original references.

The most accurately calculated numerical values of collisional excitation cross sections for H are due to

Burke, P. G., Ormonde, S., and Whitaker, W., 1967. *Proc. Phys. Soc. Lon.* **92**, 329.

(Table 3.12 is based on this reference.)

4

Calculation of Emitted Spectrum

4.1 Introduction

The radiation emitted by each element of volume in a gaseous nebula depends upon the abundances of the elements, presumably determined by the previous evolutionary history of the gas, and on the local ionization and temperature, determined by the radiation field and the abundances as described in the preceding two chapters. The most prominent spectral features are the emission lines, and many of these are the collisionally excited lines described in the preceding chapter on thermal equilibrium. The formalism developed there to calculate the cooling rate, and thus the thermal equilibrium, may be taken over unchanged to calculate the strength of these lines. If it were possible to observe all the lines in the entire spectral region from the extreme ultraviolet to the far infrared, a direct measurement would be available of the cooling rate at each observed point in the nebula. Many of the most important lines in the cooling, for instance, [O II] $\lambda\lambda 3726, 3729$ and [O III] $\lambda\lambda 4959, 5007$, are in the optical region and are easily measured. However, many other lines that are also important in the cooling, such as [O III] $2p^2\,{}^3P_0 - 2p^2\,{}^3P_1$, $\lambda 88\,\mu$, and ${}^3P_1 - {}^3P_2\,\lambda 52\,\mu$, are in the presently unobserved far infrared region, while still others, such as C IV λλ1548, 1550, are in the presently little-observed ultraviolet. For this reason, it is not yet possible to carry out in practice the direct measurement of the cooling rate of a gaseous nebula.

It should also be noted that, for historical reasons, there is a common tendency to refer to the chief emission lines of gaseous nebulae as *forbidden* lines. Actually, it is better to think of the bulk of the lines as *collisionally excited* lines, which arise from levels within a few volts of the ground level and which therefore can be excited by collisions with thermal electrons. Now in fact, in the ordinary optical region all these collisionally excited lines are forbidden lines, because in the common ions all the excited levels within a few volts of the ground level arise from the same electron configuration as the ground level itself, and thus radiative transitions are forbidden by the parity selection rule. However, just slightly below the ultraviolet cutoff of the earth's atmosphere, collisionally excited lines begin to appear that are not forbidden lines; for example, Mg II $3s\,^2S$–$2p\,^2P$ λλ2798, 2802, C IV $2s\,^2S$–$2p\,^2P$ λλ1548, 1550, and Si IV $3s\,^2S$–$3p\,^2P$ λλ1394, 1403. All these lines are calculated to be strong in the spectra of gaseous nebulae, and undoubtedly would have been observed already if it were not for the ultraviolet absorption of the earth's atmosphere.

In addition to the collisionally excited lines, the recombination lines of H I, He I, and He II are characteristic features of the spectra of gaseous nebulae. They are emitted by atoms undergoing radiative transitions in cascading down to the ground level following recombinations to excited levels. In the remainder of this chapter, these recombination emission processes will be discussed in more detail, and then the related topic of resonance-fluorescence excitation of observable lines of other elements will be considered. Finally, the continuum-emission processes, which are the bound-free and free-free analogues of the bound-bound transitions emitted in the recombination-line spectrum, will be examined.

4.2   Optical Recombination Lines

The recombination-line spectrum of H I is emitted by H atoms that have been captured into excited states and that are cascading by downward radiative transitions to the ground level. In the limit of very low density, the only processes that need be considered are captures and downward-radiative transitions. Thus, the equation of statistical equilibrium for any level nL may be written

$$N_p N_e \alpha_{nL}(T) + \sum_{n'>n}^{\infty} \sum_{L'} N_{n'L'} A_{n'L',nL} = N_{nL} \sum_{n''=1}^{n-1} \sum_{L''} A_{nL,n''L''}. \quad (4.1)$$

It is convenient to express the populations in terms of the dimensionless factors b_{nL} that measure the deviation from thermodynamic equilibrium at the local T, N_e, and N_p. Since in thermodynamic equilibrium, the Saha equation,

$$\frac{N_p N_e}{N_{1S}} = \left(\frac{2\pi m k T}{h^2}\right)^{3/2} e^{-h\nu_0/kT}, \tag{4.2}$$

and the Boltzmann equation,

$$\frac{N_{nL}}{N_{1S}} = (2L + 1)e^{-X_n/kT}, \tag{4.3}$$

apply, the population in the level nL in thermodynamic equilibrium may be written

$$N_{nL} = (2L + 1)\left(\frac{h^2}{2\pi m k T}\right)^{3/2} e^{X_n/kT} N_p N_e, \tag{4.4}$$

where

$$X_n = h\nu_0 - \chi_n = \frac{h\nu_0}{n^2} \tag{4.5}$$

is the ionization potential of the level nL. Therefore, in general, the population may be written

$$N_{nL} = b_{nL}(2L + 1)\left(\frac{h^2}{2\pi m k T}\right)^{3/2} e^{X_n/kT} N_p N_e, \tag{4.6}$$

and $b_{nL} = 1$ in thermodynamic equilibrium.

Substituting this expression in (4.1),

$$\frac{\alpha_{nL}}{(2L + 1)}\left(\frac{2\pi m k T}{h^2}\right)^{3/2} e^{-X_n/kT} + \sum_{n'>n}\sum_{L'} b_{n'L'} A_{n'L',nL}$$

$$= b_{nl} \sum_{n''=1}^{n-1}\sum_{L''} A_{nL,n''L''}, \tag{4.7}$$

it can be seen that the b_{nL} factors are independent of density as long as recombination and downward-radiative transitions are the only relevant processes. Furthermore, it can be seen that the equations (4.7) can be solved by a systematic procedure working downward in n, for if the b_{nL} are known for all $n \geqslant n_K$, then the n equations (4.7), with $L = 0, 1, \ldots, n - 1$ for

$n = n_K - 1$, each contain a single unknown b_{nL} and can be solved immediately, and so on successively downward.

It is convenient to express the solutions in terms of the cascade matrix $C_{nL,n'L'}$, which is the probability that population of nL is followed by a transition to $n'L'$ via all possible cascade routes. The cascade matrix can be generated directly from the probability matrix $P_{nL,n'L'}$, which gives the probability that population of the level nL is followed by a direct radiative transition to $n'L'$,

$$P_{nL,n'L'} = \frac{A_{nL,n'L'}}{\sum\limits_{n''=1}^{n-1}\sum\limits_{L''} A_{nL,n''L''}}, \tag{4.8}$$

which is zero unless $L' = L \pm 1$.

Hence, for $n' = n - 1$,

$$C_{nL,n-1L'} = P_{nL,n-1L'};$$

for $n' = n - 2$,

$$C_{nL,n-2L'} = P_{nL,n-2L'} + \sum_{L''=L\pm1} C_{nL,n-1L''} P_{n-1L'',n-2L'};$$

and for $n' = n - 3$,

$$C_{nL,n-3L'} = P_{nL,n-3L'} + \sum_{L''=L\pm1} (C_{nL,n-1L''} P_{n-1L'',n-3L'} + C_{nL,n-2L''} P_{n-2L'',n-3L'});$$

so that if we define

$$C_{nL,nL''} = \delta_{LL''}, \tag{4.9}$$

then, in general,

$$C_{nL,n'L'} = \sum_{n''>n'}^{n} \sum_{L''=L'\pm1} C_{nL,n''L''} P_{n''L'',n'L'}. \tag{4.10}$$

The solutions of the equilibrium equations (4.1) may be immediately written down, for the population of any level nL is fixed by the balance between recombinations to all levels $n' \geqslant n$ that lead by cascades to nL and downward radiative transitions from nL:

$$N_p N_e \sum_{n'=n}^{\infty} \sum_{L'=0}^{n'-1} \alpha_{n'L'}(T) C_{n'L',nL} = N_{nL} \sum_{n''=1}^{n-1} \sum_{L''=L\pm1} A_{nL,n''L''}. \tag{4.11}$$

It is convenient to express the results in this form, because once the cascade matrix has been calculated, it can be used to find the b_{nL} factors or the populations N_{nL} at any temperature, or even for cases in which the population occurs by other nonradiative processes, such as collisional excitation from the ground level or from an excited level. To carry out the solutions, it can be seen from (4.11) that it is necessary to fit series in n, n', L, and L' to $C_{nL,n'L'}$ and $\alpha_{nL}(T)$, and extrapolate these series as $n \to \infty$. Once the populations N_{nL} have been found, it is simple to calculate the emission coefficient in each line,

$$j_{nn'} = \frac{h\nu_{nn'}}{4\pi} \sum_{L=0}^{n-1} \sum_{L'=L\pm1} N_{nL} A_{nL,n'L'}. \qquad (4.12)$$

The situation we have been considering is commonly called case A in the theory of recombination line radiation, and assumes that all line photons emitted in the nebula escape without absorption and therefore without causing further upward transitions. Case A is thus a good approximation for gaseous nebulae that are optically thin in all H I absorption lines, but in fact such nebulae can contain only a relatively small amount of gas and are mostly too faint to be easily observed.

Nebulae that contain observable amounts of gas generally have quite large optical depths in the Lyman resonance lines of H I. This can be seen from the equation for the central line-absorption cross section,

$$a_0(Ln) = \frac{3\lambda_{n1}^3}{8\pi} \left(\frac{m_H}{2\pi k T}\right)^{1/2} A_{nP,1S}, \qquad (4.13)$$

where λ_{n1} is the wavelength of the line. Thus, at a typical temperature $T = 10,000°$K, the optical depth in $L\alpha$ is about 10^4 times the optical depth at the Lyman limit $\nu = \nu_0$ of the ionizing continuum, and an ionization-bounded nebula with $\tau_0 \approx 1$ therefore has $\tau(L\alpha) \approx 10^4$, $\tau(L\beta) \approx 10^3$, $\tau(L8) \approx 10^2$, and $\tau(L18) \approx 10$. In each scattering there is a finite probability that the Lyman-line photon will be converted to a lower-series photon plus a lower member of the Lyman series. Thus, for instance, each time an $L\beta$ photon is absorbed by an H atom, raising it to the $3\,^2P$ level, the probability that this photon is scattered is $P_{31,10} = 0.882$, while the probability that it is converted to $H\alpha$ is $P_{31,20} = 0.118$, so after nine scatterings, an average $L\beta$ photon is converted to $H\alpha$ (plus two photons in the $2\,^2S \to 1\,^2S$ continuum) and cannot escape from the nebula. Likewise, an average $L\gamma$ photon is transformed, after a relatively few scatterings, either into a $P\alpha$ photon plus an $H\alpha$ photon plus an $L\alpha$ photon, or into an $H\beta$ photon plus two photons in the $2\,^2S$–$1\,^2S$ continuum. Thus, for these large optical depths, a better approximation than case A is the opposite assumption that every Lyman-line photon is scattered many times and is converted (if $n \geqslant 3$) into lower-series photons plus either $L\alpha$ or two-continuum photons. This large optical depth approximation is called case B, and is more accurate than case A for most nebulae, though it is clear that the real situation is intermediate, and is similar to case B for the lower Lyman lines, but progresses continuously to a situation nearer case A as $n \to \infty$ and $\tau(Ln) \to \sim 1$.

Under case B conditions, any photon emitted in an $n\,^2P \to 1\,^2S$ transition is immediately absorbed nearby in the nebula, thus populating the $n\,^2P$ level in another atom. Hence, in case B, the downward-radiative transitions to $1\,^2S$ are simply omitted from consideration, and the sums in the equilibrium equations (4.1), (4.7), (4.8), and (4.11) are terminated at $n'' = n_0 = 2$ instead of $n_0 = 1$ as in case A. The detailed transition between cases A and B will be discussed in Section 4.5.

Selected numerical results from the recombination spectrum of H I are listed in Tables 4.1 and 4.2 for cases A and B, respectively. Note that, in addition to the emission coefficient $j_{42} = j_{H\beta}$ and the relative intensities of the other lines, it is also sometimes convenient to use the effective recombination coefficient, defined by

$$N_p N_e \alpha_{nn'}^{\text{eff}} = \sum_{L=0}^{n-1} \sum_{L'=L\pm1} N_{nL} A_{nL,n'L'} = \frac{4\pi j_{nn'}}{h\nu_{nn'}}. \tag{4.14}$$

For hydrogenlike ions of nuclear charge Z, all the transition probabilities $A_{nL,n'L'}$ are proportional to Z^4, so the $P_{nL,n'L'}$ and $C_{nL,n'L'}$ matrices are independent of Z. The recombination coefficients α_{nL} scale as

$$\alpha_{nL}(Z, T) = Z\,\alpha_{nL}(1, T/Z^2);$$

the effective recombination coefficients scale in this same way, and since the energies $h\nu_{nn'}$ scale as

$$\nu_{nn'}(Z) = Z^2 \nu_{nn'}(1),$$

the emission coefficient is

$$j_{nn'}(Z, T) = Z^3 j_{nn'}(1, T/Z^2). \tag{4.15}$$

Thus the calculations for H I at a particular temperature T also can be applied to He II at $T' = 4T$. In Table 4.3 some of the main features of the He II recombination-line spectrum are listed for case B, with the strongest line in the optical spectrum, $\lambda 4686$ ($n = 4 \to 3$), as the reference line.

Next let us return to the H I recombination lines and examine the effects of collisional transitions at finite nebular densities. The largest collisional cross sections involving the excited levels of H are for transitions $nL \to nL \pm 1$, which have essentially zero energy difference. Collisions with both electrons and protons can cause these angular momentum-changing

TABLE 4.1
H I *Recombination Lines (Case A)*

	T			
	2500°	5000°	10,000°	20,000°
$4\pi j_{H\beta}/N_p N_e$ (erg cm^3 sec^{-1})	2.70×10^{-25}	1.54×10^{-25}	8.30×10^{-26}	4.21×10^{-26}
$\alpha_{H\beta}^{\text{eff}}$ (cm^3 sec^{-1})	6.61×10^{-14}	3.78×10^{-14}	2.04×10^{-14}	1.03×10^{-14}
Balmer-line intensities relative to Hβ				
$j_{H\alpha}/j_{H\beta}$	3.42	3.10	2.86	2.69
$j_{H\gamma}/j_{H\beta}$	0.439	0.458	0.470	0.485
$j_{H\delta}/j_{H\beta}$	0.237	0.250	0.262	0.271
$j_{H\epsilon}/j_{H\beta}$	0.143	0.153	0.159	0.167
$j_{H8}/j_{H\beta}$	0.0957	0.102	0.107	0.112
$j_{H9}/j_{H\beta}$	0.0671	0.0717	0.0748	0.0785
$j_{H10}/j_{H\beta}$	0.0488	0.0522	0.0544	0.0571
$j_{H15}/j_{H\beta}$	0.0144	0.0155	0.0161	0.0169
$j_{H20}/j_{H\beta}$	0.0061	0.0065	0.0068	0.0071
Lyman-line intensities relative to Hβ				
$j_{L\alpha}/j_{H\beta}$	33.1	32.6	32.7	33.8
Paschen-line intensities relative to corresponding Balmer lines				
$j_{P\alpha}/j_{H\beta}$	0.684	0.562	0.466	0.394
$j_{P\beta}/j_{H\gamma}$	0.609	0.527	0.460	0.404
$j_{P\gamma}/j_{H}$	0.565	0.504	0.450	0.406
$j_{P\delta}/j_{H8}$	0.531	0.487	0.443	0.404
j_{P10}/j_{H10}	0.529	0.481	0.439	0.399
j_{P15}/j_{H15}	0.521	0.465	0.429	0.396
j_{P20}/j_{H20}	0.508	0.462	0.426	0.394

transitions, but because of the small energy difference, protons are more effective than electrons; for instance, representative values of the mean cross sections for thermal protons at $T \approx 10,000°$K are $\sigma_{2\ ^2S\to 2\ ^2P} \approx 3 \times 10^{-10}$ cm^2, $\sigma_{10\ ^2L\to 10\ ^2L\pm 1} \approx 4 \times 10^{-7}$ cm^2, and $\sigma_{20\ ^2L\to 20\ ^2L\pm 1} \approx 6 \times 10^{-6}$ cm^2. (Both of the latter are evaluated for $L \approx n/2$.) These collisional transitions must then be included in the equilibrium equations, which are modified from (4.1) to read

$$N_p N_e \alpha_{nL}(T) + \sum_{n'>n}^{\infty} \sum_{L'=L\pm 1} N_{n'L'} A_{n'L',nL} + \sum_{L'=L\pm 1} N_{nL'} N_p q_{nL',nL}$$

$$= N_{nL}\left[\sum_{n''=n_0}^{n-1} \sum_{L''=L\pm 1} A_{nL,n''L''} + \sum_{L''=L\pm 1} N_p q_{nL,nL''}\right], \quad (4.16)$$

TABLE 4.2
H I *Recombination Lines (Case B)*

	T			
	2500°	5000°	10,000°	20,000°
$4\pi j_{H\beta}/N_p N_e$ (erg cm^3 sec^{-1})	3.72×10^{-25}	2.20×10^{-25}	1.24×10^{-25}	6.62×10^{-26}
$\alpha_{H\beta}^{\text{eff}}$ (cm^3 sec^{-1})	9.07×10^{-14}	5.37×10^{-14}	3.03×10^{-14}	1.62×10^{-14}
Balmer-line intensities relative to $H\beta$				
$j_{H\alpha}/j_{H\beta}$	3.30	3.05	2.87	2.76
$j_{H\gamma}/j_{H\beta}$	0.444	0.451	0.466	0.474
$j_{H\delta}/j_{H\beta}$	0.241	0.249	0.256	0.262
$j_{H\epsilon}/j_{H\beta}$	0.147	0.153	0.158	0.162
$j_{H8}/j_{H\beta}$	0.0975	0.101	0.105	0.107
$j_{H9}/j_{H\beta}$	0.0679	0.0706	0.0730	0.0744
$j_{H10}/j_{H\beta}$	0.0491	0.0512	0.0529	0.0538
$j_{H15}/j_{H\beta}$	0.0142	0.0149	0.0154	0.0156
$j_{H20}/j_{H\beta}$	0.0059	0.0062	0.0064	0.0065
Paschen-line intensities relative to corresponding Balmer lines				
$j_{P\alpha}/j_{H\beta}$	0.528	0.427	0.352	0.293
$j_{P\beta}/j_{H\gamma}$	0.473	0.415	0.354	0.308
$j_{P\gamma}/j_H$	0.440	0.398	0.354	0.313
j_{P8}/j_{H8}	0.421	0.388	0.350	0.321
j_{P10}/j_{H10}	0.422	0.389	0.350	0.320
j_{P15}/j_{H15}	0.415	0.383	0.344	0.321
j_{P20}/j_{H20}	0.407	0.387	0.344	0.323

where $n_0 = 1$ or 2 for cases A and B, respectively, and

$$q_{nL,n'L'} \equiv q_{nL,n'L'}(T) = \int_0^\infty v \sigma_{nL \to n'L'} f(v) \, dv \qquad (4.17)$$

is the collisional transition probability per proton per unit volume. For sufficiently large proton densities, the collisional terms dominate, and because of the principle of detailed balancing, they tend to set up a thermodynamic equilibrium distribution of the various L levels within each n; that is, they tend to make

$$\frac{N_{nL}}{N_{nL'}} = \frac{(2L + 1)}{(2L' + 1)}$$

TABLE 4.3
He II *Recombination Lines (Case B)*

	T			
	5000°	10,000°	20,000°	40,000°
$4\pi j_{\lambda4686}/N_{He^{++}}N_e$ (erg cm^3 sec^{-1})	3.14×10^{-24}	1.58×10^{-24}	7.54×10^{-25}	3.48×10^{-25}
$\alpha_{\lambda4686}^{eff}$ (cm^3 sec^{-1})	7.40×10^{-13}	3.72×10^{-13}	1.77×10^{-13}	8.20×10^{-14}
Pickering-line ($n \to 4$) intensities relative to $\lambda4686$				
$j_{54}/j_{\lambda4686}$	0.295	0.274	0.256	0.237
$j_{64}/j_{\lambda4686}$	0.131	0.134	0.135	0.134
$j_{74}/j_{\lambda4686}$	0.0678	0.0734	0.0779	0.0799
$j_{84}/j_{\lambda4686}$	0.0452	0.0469	0.0506	0.0527
$j_{94}/j_{\lambda4686}$	0.0280	0.0315	0.0345	0.0364
$j_{104}/j_{\lambda4686}$	0.0198	0.0226	0.0249	0.0262
$j_{124}/j_{\lambda4686}$	0.0106	0.0124	0.0139	0.0149
$j_{154}/j_{\lambda4686}$	0.0050	0.0060	0.0069	0.0075
$j_{204}/j_{\lambda4686}$	0.0020	0.0024	0.0029	0.0031
Pfund-line ($n \to 5$) intensities relative to corresponding Pickering lines				
j_{65}/j_{64}	0.825	0.713	0.634	0.566
j_{75}/j_{74}	0.807	0.734	0.659	0.593
j_{85}/j_{84}	0.708	0.705	0.646	0.590
j_{105}/j_{104}	0.727	0.690	0.643	0.599
j_{155}/j_{154}	0.640	0.650	0.623	0.600
j_{205}/j_{204}	0.600	0.625	0.586	0.613

or

$$N_{nL} = \frac{(2L + 1)}{n^2} N_n, \tag{4.18}$$

which is equivalent to $b_{nL} = b_n$, independent of L, where

$$N_n = \sum_{L=0}^{n-1} N_{nL}$$

is the total population in the levels with the same principal quantum number n. Since the cross sections $\sigma_{nL \to nL\pm1}$ increase with increasing n, while the transition probabilities $A_{nL,n'L\pm1}$ decrease, equations (4.18) become increasingly good approximations with increasing n, and there is therefore (for any density and temperature) a level n_{cL} (for coupled angular momentum) above which they apply. For H at $T \approx 10,000°$K, this level is approximately $n_{cL} \approx 15$ at $N_p \approx 10^4$ cm^{-3}, $n_{cL} \approx 30$ at $N_p \approx 10^2$ cm^{-3}, and $n_{cL} \approx 45$ at $N_p \approx 1$ cm^{-3}.

Exactly the same type of effects occurs in the He II spectrum, because it also has the property that all the levels nL with the same n are degenerate. The He II lines are emitted in the H$^+$, He^{++} zone of a nebula, so both protons and He^{++} ions (thermal α-particles) can cause collisional, angular-momentum-changing transitions in excited levels of He$^+$. The cross sections $\sigma_{nL \to nL\pm1}$ actually are larger for the He^{++} ions than for the H$^+$ ions, and both of them must be taken into account in the He^{++} region. The principal quantum numbers above which (4.18) applies for He II at $T \approx 10,000°$K are approximately $n_{cL} \approx 22$ for $N_p \approx 10^4$ cm^{-3}, and $n_{cL} \approx 32$ for $N_p \approx 10^2$ cm^{-3}.

After the angular-momentum-changing collisions at fixed n, the next largest collisional transition rates occur for collisions in which n changes by ±1, and of these the strongest are those for which L also changes by ±1. For this type of transition, collisions with electrons are more effective than collisions with protons, and representative cross sections for thermal electrons at $T \approx 10,000°$K are of order $\sigma_{nL \to n\pm1,L\pm1} \approx 10^{-16}$ cm^2. The effects of these collisions can be incorporated into the equilibrium equations by a straightforward generalization of (4.16), and indeed, since the cross sections for collisions $\sigma_{nL \to n\pm\Delta n, L\pm1}$ decrease with increasing Δn (but not too rapidly), collisions with $\Delta n = 1, 2, 3, \ldots$ must be included. The computational work required to set up and solve the equilibrium equations numerically becomes increasingly complicated and lengthy, but is straightforward in principle. It is clear that the collisions tend to couple levels with $\Delta L = \pm1$ and small Δn, and that this coupling increases with increasing N_e (and N_p) and with increasing n. With collisions taken into account, the b_{nL} factors and the resulting emission coefficients are no longer independent of density.

A selection of calculated results for H I, including these collisional effects, is given in Table 4.4, which shows that the density dependence is rather small, so this table, together with Table 4.2, which applies in the limit $N_e \to 0$, enables the H-line emission coefficients to be evaluated over a wide range of densities and temperatures. Similarly, Table 4.5 shows calculated results for the He II recombination spectrum at finite densities and may be used in conjunction with Table 4.3, which applies in the same limit.

Exactly the same formalism can be applied to He I recombination lines, treating the singlets and triplets as completely separate systems, since all transition probabilities between them are quite small. The He I triplets, therefore, always follow case B, because downward radiative transitions to $1\,^1S$ essentially do not occur. For the singlets, case B is ordinarily a better approximation than case A for observed nebulae, though the optical depths are lower for all cases than for the corresponding lines of H by a factor

TABLE 4.4
H I Recombination Lines (Case B)

	T						
	5000°		10,000°			20,000°	
N_e (cm^{-3})	10^2	10^4	10^2	10^4	10^6	10^2	10^4
$4\pi j_{H\beta}/N_p N_e$ (erg cm^3 sec^{-1})	2.20×10^{-25}	2.22×10^{-25}	1.24×10^{-25}	1.24×10^{-25}	1.25×10^{-25}	0.658×10^{-25}	0.659×10^{-25}
$\alpha_{H\beta}^{\text{eff}}$ (cm^3 sec^{-1})	5.39×10^{-14}	5.44×10^{-14}	3.02×10^{-14}	3.04×10^{-14}	3.07×10^{-14}	1.61×10^{-14}	1.61×10^{-14}
Balmer-line intensities relative to Hβ							
$j_{H\alpha}/j_{H\beta}$	3.03	3.00	2.86	2.85	2.81	2.74	2.74
$j_{H\gamma}/j_{H\beta}$	0.459	0.460	0.469	0.469	0.471	0.476	0.476
$j_{H\delta}/j_{H\beta}$	0.252	0.253	0.259	0.259	0.262	0.264	0.264
$j_{H\epsilon}/j_{H\beta}$	0.154	0.155	0.159	0.159	0.163	0.163	0.163
$j_{H8}/j_{H\beta}$	0.102	0.102	0.105	0.105	0.110	0.107	0.107
$j_{H9}/j_{H\beta}$	0.0710	0.0714	0.0731	0.0734	0.0783	0.0746	0.0746
$j_{H10}/j_{H\beta}$	0.0516	0.0520	0.0530	0.0533	0.0588	0.0540	0.0541
$j_{H15}/j_{H\beta}$	0.0154	0.0162	0.0156	0.0162	0.0213	0.0158	0.0161
$j_{H20}/j_{H\beta}$	0.0066	0.0081	0.0066	0.0075	0.0104	0.0066	0.0071
Paschen-line intensities relative to corresponding Balmer lines							
$j_{P\alpha}/j_{H\beta}$	0.410	0.402	0.348	0.346	0.337	0.300	0.298
$j_{P\beta}/j_{H\gamma}$	0.401	0.395	0.348	0.346	0.336	0.305	0.304
$j_{P\gamma}/j_{H}$	0.395	0.390	0.348	0.346	0.334	0.308	0.307
j_{P8}/j_{H8}	0.386	0.382	0.347	0.346	0.330	0.314	0.313
j_{P10}/j_{H10}	0.379	0.376	0.347	0.346	0.326	0.318	0.317
j_{P15}/j_{H15}	0.374	0.365	0.346	0.340	0.313	0.320	0.315
j_{P20}/j_{H20}	0.372	0.346	0.346	0.328	0.308	0.320	0.310

TABLE 4.5
He II *Recombination Lines (Case B)*

	T			
	5000°	10,000°		20,000°
N_e (cm^{-3})	10^4	10^4	10^6	10^4
$4\pi j_{\lambda 4686}$(He II)$/N_{\text{He}^{++}} N_e$ (ergs cm^3 sec^{-1})	2.93×10^{-24}	1.48×10^{-24}	1.44×10^{-24}	7.16×10^{-25}
$\alpha^{\text{eff}}_{\lambda 4686}$ (cm^3 sec^{-1})	6.92×10^{-13}	3.49×10^{-13}	3.39×10^{-13}	1.69×10^{-13}
Pickering-line intensities relative to $j_{\lambda 4686}$				
$j_{54}/j_{\lambda 4686}$	0.279	0.265	0.271	0.249
$j_{64}/j_{\lambda 4686}$	0.136	0.137	0.136	0.137
$j_{74}/j_{\lambda 4686}$	0.076	0.080	0.078	0.082
$j_{84}/j_{\lambda 4686}$	0.048	0.051	0.049	0.053
$j_{94}/j_{\lambda 4686}$	0.032	0.035	0.033	0.036
$j_{104}/j_{\lambda 4686}$	0.023	0.025	0.024	0.026
$j_{124}/j_{\lambda 4686}$	0.013	0.014	0.013	0.015
$j_{154}/j_{\lambda 4686}$	0.0071	0.0074	0.0068	0.0078
$j_{204}/j_{\lambda 4686}$	0.0037	0.0036	0.0029	0.0035
Pfund-line ($n \to 5$) intensities relative to corresponding Pickering lines				
j_{65}/j_{64}	0.779	0.688	0.654	0.613
j_{75}/j_{74}	0.755	0.680	0.654	0.610
j_{85}/j_{84}	0.735	0.672	0.652	0.610
j_{105}/j_{104}	0.705	0.662	0.648	0.619
j_{155}/j_{154}	0.674	0.642	0.620	0.608
j_{205}/j_{204}	0.661	0.634	0.579	0.604

of approximately the abundance ratio. Calculated results for the strongest He I lines are summarized in Table 4.6, with λ4471 ($2\,^3P$–$4\,^3D$) as the reference line. Note that only H itself and the ions of the H isoelectronic sequence have energy levels with the same n but different L degenerate, so for He I, Table 4.6 lists the $j_{n^{(2S+1)}L, n'\,^3L'}$ rather than $j_{nn'}$ as for H. The radiative-transfer effects on the He I triplets, discussed in Section 4.6, and the collisional-excitation effects, discussed in Section 4.8, are not included in this table.

4.3   Optical Continuum Radiation

In addition to the line radiation emitted in the bound-bound transitions previously described, recombination processes also lead to the emission of rather weak continuum radiation in free-bound and free-free transitions.

TABLE 4.6
He I Recombination Spectrum (Case B)

			T				
	5000°		10,000°			20,000°	
N_e (cm^{-3})	10^2	10^4	10^2	10^4	10^6	10^2	10^4
$4\pi j_{\lambda4471}/N_{He^+}N_e$ (erg cm^3 sec^{-1})	1.16×10^{-25}	1.17×10^{-25}	6.06×10^{-26}	6.08×10^{-26}	6.16×10^{-26}	2.94×10^{-26}	2.95×10^{-26}
$\alpha^{eff}_{\lambda4471}$ (cm^3 sec^{-1})	2.61×10^{-14}	2.64×10^{-14}	1.36×10^{-14}	1.37×10^{-14}	1.38×10^{-14}	0.662×10^{-14}	0.663×10^{-14}
			Triplet lines				
$j_{\lambda5876}/j_{\lambda4471}$	3.02	3.01	2.76	2.76	2.73	2.58	2.58
$j_{\lambda4026}/j_{\lambda4471}$	0.458	0.459	0.474	0.474	0.476	0.487	0.487
$j_{\lambda3820}/j_{\lambda4471}$	0.251	0.251	0.264	0.264	0.265	0.274	0.274
$j_{\lambda7065}/j_{\lambda4471}$	0.244	0.243	0.330	0.328	0.325	0.478	0.477
$j_{\lambda10830}/j_{\lambda4471}$	3.98	3.96	4.42	4.42	4.41	5.02	5.01
$j_{\lambda3889}/j_{\lambda4471}$	1.89	1.90	2.26	2.26	2.27	2.79	2.79
$j_{\lambda3187}/j_{\lambda4471}$	0.748	0.747	0.916	0.917	0.920	1.16	1.16
			Singlet lines				
$j_{\lambda6678}/j_{\lambda4471}$	—	0.867	—	0.791	0.780	—	0.731
$j_{\lambda4922}/j_{\lambda4471}$	—	0.276	—	0.274	0.274	—	0.271
$j_{\lambda5016}/j_{\lambda4471}$	—	0.512	—	0.588	0.590	—	0.689
$j_{\lambda3965}/j_{\lambda4471}$	—	0.199	—	0.234	0.235	—	0.279

Because hydrogen is the most abundant element, the H I continuum, emitted in the recombination of protons with electrons, is the strongest, and the He II continuum may also be significant if He is mostly doubly ionized, while the He I continuum is always weaker. In the ordinary optical region the free-bound continua are stronger, but in the infrared and radio regions the free-free continuum dominates. In addition, there is a continuum resulting from the two-photon decay of the $2\,^2S$ level of H, which is populated by recombinations and subsequent downward cascading. In this section we shall examine each of these sources of continuous radiation.

The H I free-bound continuum radiation at frequency ν results from recombinations of free electrons with velocity v to levels with principal quantum number $n \geqslant n_1$, where

$$h\nu = \frac{1}{2}mv^2 + X_n \qquad (4.19)$$

and

$$h\nu \geqslant X_{n_1} = \frac{h\nu_0}{n_1^2}; \qquad (4.20)$$

its emission coefficient per unit frequency interval per unit solid angle per unit time per unit volume is therefore

$$j_\nu = \frac{1}{4\pi} N_p N_e \sum_{n=n_1}^{\infty} \sum_{L=0}^{n-1} v\sigma_{nL}(\mathrm{H}^0, v)f(v)h\nu\,\frac{dv}{d\nu}. \qquad (4.21)$$

The recombination cross sections $\sigma_{nL}(\mathrm{H}^0, v)$ can be calculated from the photoionization cross sections $a_\nu(\mathrm{H}, nL)$ by the Milne relation, as shown in Appendix 1.

The free-free (or bremsstrahlung) continuum emitted by free electrons accelerated in Coulomb collisions with positive ions (which are mostly H^+, He^+, or He^{++} in nebulae) of charge Z has emission coefficient

$$j_\nu = \frac{1}{4\pi} N_+ N_e \frac{32Z^2 e^4 h}{3m^2 c^3}\left(\frac{\pi h\nu_0}{3kT}\right)^{1/2} e^{-h\nu/kT} g_{ff}(T, Z, \nu), \qquad (4.22)$$

where $g_{ff}(T, Z, \nu)$ is a mean Gaunt factor. Thus the emission coefficient for the H I recombination continuum, including both bound-free and free-free contributions, may be written

$$j_\nu(HI) = \frac{1}{4\pi} N_p N_e \gamma_\nu(\mathrm{H}^0, T). \qquad (4.23)$$

Numerical values for γ_ν, as calculated from equations (4.21) and (4.22), are given in Table 4.7. Likewise, the contributions to the continuum emission coefficient from He I and He II may be written

$$j_\nu(\text{He I}) = \frac{1}{4\pi} N_{\text{He}^+} N_e \gamma_\nu(\text{He}^0, T),$$

$$j_\nu(\text{He II}) = \frac{1}{4\pi} N_{\text{He}^{++}} N_e \gamma_\nu(\text{He}^+, T), \qquad (4.24)$$

and numerical values of the γ_ν are listed in Tables 4.8 and 4.9. The calculation for He II is exactly analogous to that for H I, while for He I the only complication is that there is no L degeneracy and equations (4.19), (4.20), and (4.21) must be appropriately generalized. Figure 4.1 shows these calculated values of γ_ν graphically, and also shows the large discontinuities at the ionization potentials of the various excited levels. Note that for a typical He abundance of approximately 10 percent of the H abundance, if the He is mostly doubly ionized, then the He II contribution to the continuum is roughly comparable to that of H I, but if the He is mostly singly ionized, the He I contribution to the continuum is only about 10 percent of the H I contribution.

An additional important source of continuum emission in nebulae is the two-photon decay of the $2\,{}^2S$ level of H I, which is populated by direct recombinations and by cascades following recombinations to higher levels. The transition probability for this two-photon decay is $A_{2\,{}^2S,\,1\,{}^2S} = 8.23$ sec^{-1}, and the sum of the energies of the two photons is $h\nu' + h\nu'' = h\nu_{12} = h\nu(L\alpha) = (3/4)h\nu_0$. The probability distribution of the emitted photons is

TABLE 4.7
H I *Continuous-Emission Coefficient* $\gamma_\nu(\text{H, T})$*

λ (A)	ν (10^{14} Hz)	T			
		5000°	10,000°	15,000°	20,000°
10,000	2.998	6.23	5.86	5.35	4.98
8204+	3.654−	3.30	4.31	4.35	4.25
8204−	3.654+	25.36	11.54	8.31	6.81
7000	4.283	13.51	8.67	6.87	5.90
5696	5.263	5.17	5.52	5.09	4.70
4500	6.662	1.388	2.88	3.29	3.40
4000	7.495	0.653	1.946	2.54	2.79
3646+	8.224−	0.343	1.380	2.017	2.35
3646−	8.224+	71.7	24.8	14.89	10.65
3122	9.603	17.83	13.19	9.81	7.84
2600	11.530	2.857	5.39	5.43	5.07

* In 10^{-40} erg cm^3 sec^{-1} Hz^{-1}.

TABLE 4.8
He I *Continuous-Emission Coefficient* γ_ν(He, T)*

λ (A)	ν (10^{14} Hz)	T			
		5000°	10,000°	15,000°	20,000°
10,000	2.998	6.23	5.86	5.36	4.98
8268+	3.626−	3.40	4.36	4.38	4.27
8268−	3.626+	5.66	5.10	4.79	4.54
8197+	3.657−	5.52	5.04	4.74	4.50
8197−	3.657+	8.05	5.88	5.20	4.81
8195+	3.658−	8.05	5.88	5.20	4.81
8195−	3.658+	13.06	8.50	6.65	5.73
7849+	3.819−	14.14	8.08	6.44	5.59
7849−	3.819+	21.21	10.61	7.86	6.52
7440+	4.029−	18.04	10.00	7.58	6.35
7440−	4.029+	18.58	10.18	7.68	6.41
6636+	4.518−	12.41	8.66	6.94	5.96
6636−	4.518+	13.71	9.09	7.17	6.11
5696	5.263	7.07	6.76	5.91	5.30
4500	6.662	1.775	3.31	3.63	3.66
4000	7.495	0.759	2.07	2.62	2.85
3680+	8.147−	0.406	1.45	2.06	2.37
3680−	8.147+	13.37	5.69	4.38	3.87
3422+	8.761−	7.22	4.32	3.65	3.37
3422−	8.761+	64.1	23.0	13.88	9.96
3122+	9.603−	27.6	15.87	10.96	8.44
3122−	9.603+	32.5	17.45	11.83	9.00
2600	11.530	5.0	7.14	6.60	5.91

* In 10^{-40} erg cm³ sec⁻¹ Hz⁻¹.

TABLE 4.9
He II *Continuous-Emission Coefficient* γ_ν(He⁺, T)*

λ (A)	ν (10^{14} Hz)	T			
		5000°	10,000°	15,000°	20,000°
10,000	2.998	40.9	28.8	24.0	21.2
8200+	3.654−	21.4	21.1	19.48	18.12
8200−	3.654+	67.3	36.2	27.8	23.4
7000	4.283	35.8	27.0	22.8	20.2
5694+	5.263−	13.69	17.05	16.75	16.05
5694−	5.263+	92.3	42.8	29.9	25.2
4500	6.662	22.7	22.3	19.99	18.20
4000	7.495	10.16	15.07	15.39	14.98
3644+	8.224−	5.14	10.69	12.24	12.63
3644−	8.224+	156.5	60.4	39.6	30.2
3122	9.603	38.6	31.7	25.8	22.0
2600	11.530	6.15	12.79	14.07	14.05

* In 10^{-40} erg cm³ sec⁻¹ Hz⁻¹.

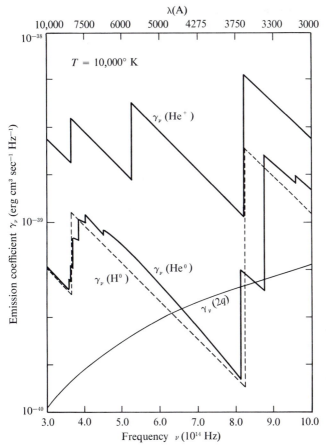

FIGURE 4.1

Frequency variation of continuous-emission coefficients $\gamma_\nu(H^0)$, $\gamma_\nu(He^0)$, $\gamma_\nu(He^+)$ and $\gamma_\nu(2q)$ in the low-density limit $N_e \to 0$, all at $T = 10,000^\circ K$.

therefore symmetric around the frequency $(1/2)\nu_{12} = 1.23 \times 10^{15}$ sec^{-1}, corresponding to $\lambda = 2431$ A. The emission coefficient in this two-photon continuum may be written

$$j_\nu(2q) = \frac{1}{4\pi} N_{2\,^2S} A_{2\,^2S,\,1\,^2S} 2h\nu P(y), \qquad (4.25)$$

where $P(y)\,dy$ is the normalized probability per decay that one photon is emitted in the range of frequencies $y\nu_{12}$ to $(y + dy)\nu_{12}$.

To express this two-photon continuum-emission coefficient in terms of the proton and electron density, it is necessary to calculate the equilibrium population of $N_{2\,^2S}$ in terms of these quantities. In sufficiently low-density nebulae, two-photon decay is the only mechanism that depopulates $2\,^2S$, and the equilibrium is given by

$$N_p N_e \alpha_{2\,^2S}^{\text{eff}}(H^0, T) = N_{2\,^2S} A_{2\,^2S,\,1\,^2S}, \tag{4.26}$$

where $\alpha_{2\,^2S}^{\text{eff}}$ is the effective recombination coefficient for populating $2\,^2S$ by direct recombinations and by recombinations to higher levels followed by cascades to $2\,^2S$. However, at finite densities, angular momentum-changing collisions of protons and electrons with H atoms in the $2\,^2S$ level shift the atoms to $2\,^2P$ and thus remove them from $2\,^2S$. The protons are more effective than electrons, whose effects, however, are not completely negligible, as the numerical values of the collisonal transition rates per $2\,^2S$ atom, listed in Table 4.10, show. With these collisional processes taken into account, the equilibrium population in $2\,^2S$ is given by

$$N_p N_e \alpha_{2\,^2S}^{\text{eff}}(H^0, T) = N_{2\,^2S}\{A_{2\,^2S,\,1\,^2S} + N_p q_{2\,^2S,\,2\,^2P}^p + N_e q_{2\,^2S,\,2\,^2P}^e\} \tag{4.27}$$

From Table 4.10, it can be seen that collisional de-excitation of $2\,^2S$ via $2\,^2P$ is more important than two-photon decay for $N_p \gtrsim 10^4\,\text{cm}^{-3}$, so at densities approaching this value, equation (4.27) must be used instead of equation (4.26). Thus, combining equations (4.25) and (4.27), the emission coefficient can be written

$$j_\nu(2q) = \frac{1}{4\pi} N_p N_e \gamma_\nu(2q), \tag{4.28}$$

TABLE 4.10
Collisional Transition Rates* for H I 2 ^2S, 2 ^2P

	T	
	10,000°	20,000°
	Protons	
$q_{2\,^2S,\,2\,^2P_{1/2}}^p$	2.51×10^{-4}	2.08×10^{-4}
$q_{2\,^2S,\,2\,^2P_{3/2}}^p$	2.23×10^{-4}	2.19×10^{-4}
	Electrons	
$q_{2\,^2S,\,2\,^2P_{1/2}}^e$	0.22×10^{-4}	0.17×10^{-4}
$q_{2\,^2S,\,2\,^2P_{3/2}}^e$	0.35×10^{-4}	0.27×10^{-4}
	Total	
$q_{2\,^2S,\,2\,^2P}$	5.37×10^{-4}	4.75×10^{-4}

* All values of q are in units $\text{cm}^3\,\text{sec}^{-1}$.

TABLE 4.11
Effective Recombination Coefficient to H $(2\,^2S)$

T	$\alpha_{2\,^2S}^{\text{eff}}$ (cm³ sec⁻¹)
5000°	1.38 $\times$ 10⁻¹³
10,000°	0.838 $\times$ 10⁻¹³
15,000°	0.625 $\times$ 10⁻¹³
20,000°	0.506 $\times$ 10⁻¹³

where

$$\gamma_\nu(2q) = \frac{\alpha_{2\,^2S}^{\text{eff}}(H^0, T)g_\nu}{1 + \left(\dfrac{N_p q_{2\,^2S,\,2\,^2P}^p + N_e q_{2\,^2S,\,2\,^2P}^e}{A_{2\,^2S,\,1\,^2S}}\right)}. \tag{4.29}$$

The quantity $\alpha_{2\,^2S}^{\text{eff}}$ is tabulated in Table 4.11 and g_ν is tabulated in Table 4.12. The two-photon continuum is also plotted in Figure 4.2 for $T = 10,000°$K and in the low-density limit $N_p \approx N_e \ll 10^4$ cm⁻³. It can be seen that this continuum is quite significant in comparison with the H I continua, particularly just above the Balmer limit at $\lambda 3646$ A.

4.4 Radio-Frequency Continuum and Line Radiation

The line and continuous spectra described in the preceding two sections extend to arbitrarily low frequency, and in fact give rise to observable features in the radio-frequency spectral region. Though this "thermal"

TABLE 4.12
Spectral Distribution of H I *Two-photon Emission*

λ (A)	ν (10¹⁴ Hz)	g_ν (10⁻²⁷ erg Hz⁻¹)
∞	0.0	0.0
24,313	1.23	0.303
12,157	2.47	0.978
8104	3.70	1.836
6078	4.93	2.78
4863	6.17	3.78
4052	7.40	4.80
3473	8.64	5.80
3039	9.87	6.78
2701	11.10	7.74
2431	12.34	8.62

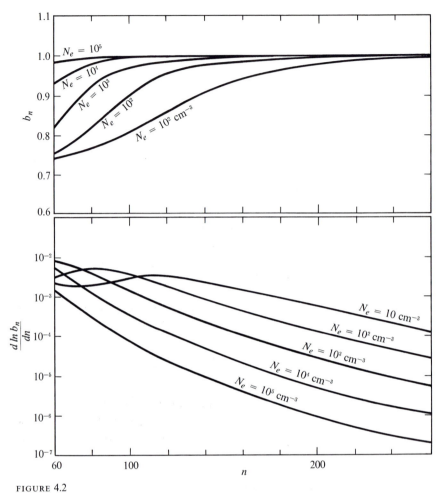

FIGURE 4.2

Dependence of b_n and $\dfrac{d \ln b_n}{dn}$ on n at various densities; $T = 10,000°$K.

radio-frequency radiation is a natural extension of the optical line and continuous spectra, it is somewhat different in detail, because in the radio-frequency region, $h\nu \ll kT$; and stimulated emission, which is proportional to $e^{-h\nu/kT}$, is therefore much more important in that region than in the ordinary optical region. We shall examine the continuous spectrum first, and then the recombination-line spectrum.

In the radio-frequency region, the continuum is due to free-free emission, and the emission coefficient is given by the same equation (4.22) that applies in the optical region. However, in the radio-frequency region, the Gaunt factor $g_{ff}(T, Z, \nu) \neq 1$, as in the optical region, but rather

$$g_{ff}(T, Z, \nu) = \frac{\sqrt{3}}{\pi} \left\{ \ln \left(\frac{8k^3 T^3}{\pi^2 Z^2 e^4 m \nu^2} \right)^{1/2} - \frac{5\gamma}{2} \right\}, \qquad (4.30)$$

where $\gamma = 0.577$ is Euler's constant. Numerically, this is approximately

$$g_{ff}(T, Z, \nu) = \frac{\sqrt{3}}{\pi} \left(\ln \frac{T^{3/2}}{Z\nu} + 17.7 \right),$$

with T in °K and ν in Hz, and thus at $T \approx 10{,}000°\mathrm{K}$, $\nu \approx 10^3$ MHz, $g_{ff} \approx 10$.

The free-free effective absorption coefficient can then be found from Kirchoff's law, and is

$$\kappa_\nu = N_+ N_e \frac{16\pi^2 Z^2 e^6}{(6\pi m k T)^{3/2} \nu^2 c} g_{ff} \qquad (4.31)$$

per unit length. Note that this effective absorption coefficient is the difference between the true absorption coefficient and the stimulated emission coefficient, since the stimulated emission of a photon is exactly equivalent to a negative absorption process; in the radio-frequency region ($h\nu \ll kT$), the stimulated emissions very nearly balance the true absorptions and the correction for stimulated emission, $(1 - e^{-h\nu/kT}) \approx h\nu/kT \ll 1$.

Substituting numerical values and fitting powers to the weak temperature and frequency dependence of g_{ff},

$$\tau = \int \kappa_\nu \, ds$$

$$= 8.24 \times 10^{-2} \, T^{-1.35} \nu^{-2.1} \int N_+ N_e \, ds$$

$$= 8.24 \times 10^{-2} \, T^{-1.35} \nu^{-2.1} E. \qquad (4.32)$$

It must be noted that in this formula T is to be measured in °K, ν in GHz, and E, the so-called emission measure, in cm^{-6} pc. It can be seen from equations (4.31) or (4.32) that, at sufficiently low frequency, all nebulae become optically thick; for example, an H II region with $N_e \approx N_p \approx 10^2\ \mathrm{cm}^{-3}$ and a diameter 10 pc has $\tau_\nu \approx 1$ at $\nu \approx 200$ MHz, and a planetary nebula with $N_e \approx N_p \approx 3 \times 10^3\ \mathrm{cm}^{-3}$ and a diameter 0.1 pc has $\tau_\nu \approx 1$ at $\nu \approx 600$ MHz. Thus, in fact, many nebulae are optically thick at observable low frequencies and optically thin at observable high frequencies. The equation of radiative transfer,

$$\frac{dI_\nu}{ds} = -\kappa_\nu I_\nu + j_\nu \qquad (4.33)$$

or

$$\frac{dI_\nu}{d\tau_\nu} = -I_\nu + \frac{j_\nu}{\kappa_\nu} = -I_\nu + B_\nu(T), \tag{4.34}$$

has the solution for no incident radiation

$$I_\nu = \int_0^{\tau_\nu} B_\nu(T)e^{-\tau_\nu}\,d\tau_\nu. \tag{4.35}$$

In the radio-frequency region,

$$B_\nu(T) = \frac{2h\nu^3}{c^2}\frac{1}{e^{h\nu/kT}-1} \approx \frac{2\nu^2 kT}{c^2} \tag{4.36}$$

is proportional to T, so it is conventional in radio astronomy to measure intensity in terms of brightness temperature, defined by $T_{b\nu} = c^2 I_\nu/2\nu^2$. Hence (4.35) can be rewritten

$$T_{b\nu} = \int_0^{\tau_\nu} Te^{-\tau_\nu}\,d\tau_\nu, \tag{4.37}$$

and for an isothermal nebula, this becomes

$$T_{b\nu} = T(1 - e^{-\tau_\nu})\begin{cases} \to T\tau_\nu & \text{as} & \tau_\nu \to 0 \\ \to T & \text{as} & \tau_\nu \to \infty \end{cases}.$$

Thus the radio-frequency continuum has a spectrum in which $T_{b\nu}$ varies approximately as ν^{-2} at high frequency and is independent of ν at low frequency.

The H I recombination lines of very high n also fall in the radio-frequency spectral region and have been observed in many gaseous nebulae. Some specific examples of observed lines are H 109α (the transition with $\Delta n = 1$ from $n = 110$ to $n = 109$) at $\nu = 5008.9$ MHz, $\lambda = 5.99$ cm, H 137β (the transition with $\Delta n = 2$ from $n = 139$ to $n = 137$) at $\nu = 5005.0$ MHz, $\lambda = 6.00$ cm, and so on. The emission coefficients in these radio recombination lines may be calculated from equations similar to those described in Section 4.2 for the shorter wavelength optical recombination lines. For all lines observed in the radio-frequency region, $n > n_{cL}$ defined there, so that at a fixed n, $N_{nL} \propto (2L + 1)$, and only the populations N_n need be considered. One additional process, in addition to those described in Section 4.2, must also be taken into account, namely, collisional ionization of levels with large n and its inverse process, three-body recombination,

$$H^0(n) + e \rightleftarrows H^+ + e + e.$$

The rate of collisional ionization per unit volume per unit time from level n may be written

$$N_n N_e \overline{v \sigma_{\text{ionization}}(n)} = N_n N_e q_{n,i}(T), \tag{4.38}$$

while the rate of three-body recombination per unit time per unit volume may be written $N_p N_e^2 \phi_n(T)$, and from the principle of detailed balancing,

$$\phi_n(T) = n^2 \left(\frac{h^2}{2\pi m k T} \right)^{3/2} e^{X_n/kT} q_{n,i}(T). \tag{4.39}$$

Thus the equilibrium equation that is analogous to equation (4.16) becomes, at high n,

$$N_p N_e [\alpha_n(T) + N_e \phi_n(T)] + \sum_{n' > n}^{\infty} N_{n'} A_{n',n} + \sum_{n=n_0}^{\infty} N_{n'} N_e q_{n',n}$$

$$= N_n \left[\sum_{n'=n_0}^{n-1} A_{n,n'} + \sum_{n'=n_0}^{\infty} N_e q_{n,n'}(T) + N_e q_{n,i}(T) \right], \tag{4.40}$$

where

$$A_{n,n'} = \frac{1}{n^2} \sum_{L,L'} (2L + 1) A_{nL,n'L'} \tag{4.41}$$

is the mean transition probability averaged over all the L levels of the upper principal quantum number. These equations can be expressed in terms of b_n instead of N_n, and the solutions can be found numerically by standard matrix inversion techniques. Note that since the coefficients b_n have been defined with respect to thermodynamic equilibrium at the local T, N_e, and N_p, the coefficient b for the free electrons is identically unity, and therefore, $b_n \to 1$ as $n \to \infty$. Some calculated values of b_n for $T = 10,000°$K and various N_e are plotted in Figure 4.2, which shows that the increasing importance of collisional transitions as N_e increases makes $b_n \approx 1$ at lower and lower n.

To calculate the emission in a particular recombination line, it is again necessary to solve the equation of transfer, taking account of the effects of stimulated emission. In this case, for an $n, \Delta n$ line between the upper level $m = n + \Delta n$ and the lower level n, if $k_{\nu l}$ is the true line-absorption coefficient, then the line-absorption coefficient, corrected for stimulated emission, to be used in the equation of transfer is

$$k_{\nu L} = k_{\nu l} \left(1 - \frac{b_m}{b_n} e^{-h\nu/kT} \right), \tag{4.42}$$

since it is the net difference between the rates of upward absorption processes and of downward-induced emissions. Expanding in a power series, this becomes

$$k_{\nu L} = k_{\nu l} \left(\frac{b_m}{b_n} \frac{h\nu}{kT} - \frac{d \ln b_n}{dn} \Delta n \right). \tag{4.43}$$

Since $b_m/b_n \approx 1$ and $h\nu \ll kT$, the line-absorption coefficient can become negative, implying positive maser action, if $(d \ln b_n)/dn$ is sufficiently large. Calculated values of this derivative are therefore also shown in Figure 4.2. Since, for typical observed lines $h\nu/kT \approx 10^{-5}$, it can be seen from this figure that the maser effect in fact often is quite important. We shall again use these concepts and expressions to calculate the strengths of the radio-frequency recombination lines in Chapter 5.

4.5 Radiative Transfer Effects in H I

For the bulk of the emission lines observed in nebulae there is no radiative-transfer problem; in most lines the nebulae are optically thin, and any line photon emitted simply escapes. However, in some lines, particularly the resonance lines of abundant atoms, the optical depths are appreciable, and scattering and absorption must be taken into account in calculating the expected line strengths. Two extreme assumptions, case A, a nebula with vanishing optical thickness in all the H I Lyman lines, and case B, a nebula with large optical depths in all the Lyman lines, have already been discussed in Section 4.2; and although in these two cases the detailed radiative-transfer solution is not required, it is clear that in the intermediate cases a more sophisticated treatment is necessary. Other radiative transfer problems arise in connection with the He I triplets, the conversion of He II $L\alpha$ into observable O III line radiation by the Bowen resonance-fluorescence process, and fluorescence excitation of other lines by stellar continuum radiation. In this section some of the general ideas of the escape of line photons from nebulae will be discussed in the context of the H I Lyman and Balmer lines, and then in succeeding sections these same ideas will be applied to the other problems mentioned.

In a static nebula the only line-broadening mechanisms are thermal Doppler broadening and radiative damping, and in the cores of the lines the line-absorption coefficient has the Doppler form

$$k_{\nu l} = k_{0l} e^{-(\Delta\nu/\Delta\nu_D)^2} = k_{0l} e^{-x^2}, \tag{4.44}$$

where

$$k_{0l} = \frac{\lambda^2}{8\pi^{3/2}} \frac{\omega_j}{\omega_i} \frac{A_{j,i}}{\Delta\nu_D} = \frac{\sqrt{\pi} \, e^2 f_{ij}}{mc \, \Delta\nu_D} \tag{4.45}$$

is the line-absorption cross section per atom at the center of the line,

$$\Delta\nu_D = \sqrt{\frac{2kT}{Mc^2}}\,\nu_0$$

is the thermal Doppler width, and f_{ij} is the f-value between the lower and upper levels i, j. Small-scale turbulence can presumably be taken into account as a further source of broadening of the line-absorption coefficient, while larger-scale turbulence and expansion of the nebula must be treated by considering the frequency shift between the emitting and absorbing volumes.

In a static nebula, a photon emitted at a particular point in a particular direction and with a particular normalized frequency x from the center of the line has a probability $e^{-\tau_x}$ of escaping from the nebula without further scattering and absorption, where τ_x is the optical depth from the point to the edge of the nebula in this direction and at this frequency. Averaging over all directions gives the mean escape probability from this point and at this frequency, and further averaging over the frequency profile of the emission coefficient gives the mean escape probability from the point. For all the forbidden lines and for most of the other lines, the optical depths are so small in every direction, even at the center of the line, that the mean escape probabilities from all points are essentially unity. However, for lines of larger optical depth we must examine the probability of escape quantitatively.

Consider an idealized spherical homogeneous nebula, with optical radius in the center of a line τ_{0l}. So long as $\tau_{0l} \lesssim 10^4$, only the Doppler core of the line-absorption cross section need be considered. The photons are emitted with the same Doppler profile, and the mean probability of escape must therefore be averaged over this profile. If, at a particular normalized frequency x, the optical radius of the nebula is τ_x, the mean probability of escape averaged over all directions and volumes is

$$p(\tau_x) = \frac{3}{4\tau_x}\left[1 - \frac{1}{2\tau_x^2} + \left(\frac{1}{\tau_x} + \frac{1}{2\tau_x^2}\right)e^{-2\tau_x},\right] \tag{4.46}$$

as shown in Appendix 2. Averaging over the Doppler profile, the mean escape probability for a photon emitted in the line is

$$\epsilon(\tau_{0l}) = \frac{1}{\sqrt{\pi}}\int_{-\infty}^{\infty} p(\tau_x)e^{-x^2}\,dx, \tag{4.47}$$

where τ_{0l} is the optical radius in the center of the line. This integral must be evaluated numerically, but for optical radii ($\tau_{0l} \leqslant 50$) that are not too large, the results can be fitted fairly accurately with $\epsilon(\tau_{0l}) = 1.72/(\tau_{0l} + 1.72)$.

If we consider a particular Lyman line Ln, photons emitted in this line that do not escape from the nebula are absorbed, and each absorption process represents an excitation of the $n\,^2P$ level of H I. This excited level very quickly undergoes a radiative decay, and the result is either resonance scattering or resonance fluorescence excitation of another H I line. If the photon emitted when the $n\,^2P$ level decays is a $1\,^2S$–$n\,^2P$ transition, the process is resonance scattering of an Ln photon; if it is emitted in the $2\,^2S$–$n\,^2P$ transition, the process is conversion of Ln into Hn plus excitation of $2\,^2S$, leading to emission of two photons in the continuum; if it is emitted in the $3\,^2S$–$n\,^2P$ transition, the process is conversion of Ln into Pn plus excitation of $3\,^2S$, leading to emission of Hα plus Lα, and so on. The probabilities of each of these processes may be found directly from the probability matrices $C_{nL,n'L}$ and $P_{nL,n'L'}$ defined in Section 4.2. If we define $P_n(Lm)$ and $P_n(Hm)$ as the probabilities that absorption of an Ln photon results in emission of an Lm photon and of an Hm photon, respectively, then

$$P_n(Lm) = C_{n1,m1}P_{m1,10} \tag{4.48}$$

and

$$P_n(Hm) = C_{n1,m0}P_{m0,21} + C_{n1,m1}P_{m1,20} + C_{n1,m2}P_{m2,21}. \tag{4.49}$$

We can now use these probabilities to calculate the emergent Lyman-line spectrum emitted from a model nebula. It is easiest to work in terms of numbers of photons emitted, and if we write R_n for the total number of Ln photons generated in the nebula per unit time by recombination and subsequent cascading, and A_n as the total number of Ln photons absorbed in the nebula per unit time, then J_n, the total number of Ln photons emitted in the nebula per unit time, is the sum of the contributions from recombination and from resonance fluorescence plus scattering:

$$J_n = R_n + \sum_{m=n}^{\infty} A_m P_m(Ln). \tag{4.50}$$

Since each Ln photon emitted has a probability ϵ_n of escaping, the total number of Ln photons escaping the nebula per unit time is

$$E_n = \epsilon_n J_n = \epsilon_n \left[R_n + \sum_{m=n}^{\infty} A_m P_m(Ln) \right]. \tag{4.51}$$

Finally, in a steady state the number of Ln photons emitted per unit time is equal to the sum of the numbers absorbed and escaping per unit time,

$$J_n = A_n + E_n = A_n + \epsilon_n J_n. \tag{4.52}$$

Thus, eliminating J_n between (4.51) and (4.52),

$$A_n = (1 - \epsilon_n)\left[R_n + \sum_{m=n}^{\infty} A_m P_m (Ln)\right],\qquad (4.53)$$

and since the R_n and $P_m (Ln)$ are known from the recombination theory and the ϵ_n is known from the radiative-transfer theory, equations (4.52) can be solved for the A_n by a systematic procedure, working downward from the highest n at which ϵ_n differs appreciably from unity. Then from these values of R_n, the E_n may be calculated from (4.51), giving the emergent Lyman-line spectrum. This is perhaps less interesting from an observational point of view than the Balmer-line spectrum, which we shall next investigate.

Let us write S_n for the number of Hn photons generated in the nebula per unit time by recombination and subsequent cascading, and suppose that there is no absorption of these Balmer-line photons, so that K_n, the total number of Hn photons emitted in the nebula per unit time, is the sum of contributions from recombination and from resonance fluorescence due to Lyman-line photons,

$$K_n = S_n + \sum_{m=n}^{\infty} A_m P_m (Hn).$$

Then, since the S_n and $P_m (Hn)$ are known from the recombination theory and the A_m is known from the previous Lyman-line solution, the K_n can be calculated immediately to obtain the emergent Balmer-line spectrum. Note that R_n, S_n, J_n, K_n, and A_n are proportional to the total number of photons; the equations are linear in these quantities; and the entire calcula- tion can therefore be normalized to any particular S_n, for instance, to S_4, the number of Hβ photons that would be emitted if there were no absorption effects. The results, in the form of calculated ratios of Hα/Hβ and Hβ/Hγ intensities, are shown in Figure 4.3 as a function of $\tau_{0l}(L\alpha)$, the optical radius of the nebula at the center of $L\alpha$; and the transition from case A ($\tau_{0l} \to 0$) to case B ($\tau_{0l} \to \infty$) can be seen clearly.

Although in most nebulae, the optical depths in the Balmer lines are small, it is possible to imagine cases in which the density $N_{H^0}(2\,^2S)$ is sufficiently high so that some self-absorption does occur in these lines. The optical depths in the Balmer lines can again be calculated from equation (4.45), and since they are all proportional to $N_{H^0}(2\,^2S)$, the radiative-transfer problem is now a function of two variables, $\tau_{0l}(L\alpha)$, giving the optical radius in the Lyman lines, and another, say, $\tau_{0l}(H\alpha)$, giving the optical radius in the Balmer lines. Although the equations are much more complicated, since now Balmer-line photons may be scattered or converted into Lyman-line photons and vice versa, there is no new effect in principle, and the same

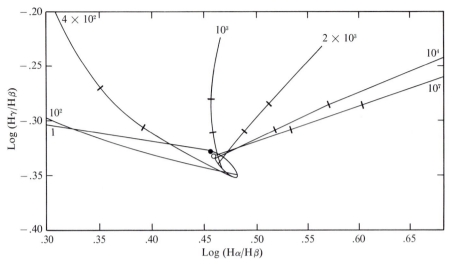

FIGURE 4.3

Radiative-transfer effects caused by finite optical depths in Lyman and Balmer lines. Ratios of total emitted fluxes $H\alpha/H\beta$ and $H\gamma/H\beta$ are shown for homogeneous static isothermal model nebulae at $T = 10{,}000^{\circ}K$. Each line connects a series of models with the same $\tau_{0l}(L\alpha)$, given at the end of the line; along it $\tau_{0l}(H\alpha) = 0$ on the locus of points shown by the loop near the center, and $\tau_{0l}(H\alpha) = 5$ and 10 at the two points along each line indicated by bars for $\tau_{0l}(L\alpha) \geqslant 4 \times 10^2$.

general type of formulation developed previously for the Lyman-line absorption can still be used. We shall not examine the details here, but will simply discuss physically the calculated results shown in Figure 4.3. For $\tau_{0l}(H\alpha) = 0$, the first effect of increasing $\tau_{0l}(L\alpha)$ is that $L\beta$ is converted into $H\alpha$ plus the two-photon continuum. This increases the $H\alpha/H\beta$ ratio of the escaping photons, corresponding to a move of the representative point to the right in Figure 4.3. However, for slightly larger $\tau_{0l}(L\alpha)$, $L\gamma$ photons are also converted into $P\alpha$, $H\alpha$, $H\beta$, $L\alpha$, and two-photon continuum photons, and since the main effect is to increase the strength of $H\beta$, this corresponds to a move downward and to the left in Figure 4.3. For still larger $\tau_{0l}(L\alpha)$, as still higher Ln photons are converted, $H\gamma$ is also strengthened, and the calculation, which takes account of all these effects, shows that the representative point describes the small loop of Figure 4.3 as the conditions change from case A to case B. For large $\tau_{0l}(L\alpha)$, the effect of increasing $\tau_{0l}(H\alpha)$ is that, although $H\alpha$ is merely scattered (because any $L\beta$ photons it forms are quickly absorbed and converted back to $H\alpha$), $H\beta$ is absorbed and converted to $H\alpha$ plus $P\alpha$. This increases $H\alpha/H\beta$ and decreases $H\beta/H\gamma$, as shown quantitatively in Figure 4.3.

4.6   Radiative Transfer Effects in He I

The recombination radiation of He I singlets is very similar to that of H I, and case *B* is a good approximation for the He I Lyman lines. However, the recombination radiation of the He I triplets is modified by the fact that the He^0 $2\,^3S$ term is considerably more metastable than H^0 $2\,^2S$, and as a result self-absorption effects are quite important (as is collisional excitation from $2\,^3S$, to be discussed later). As the energy-level diagram of Figure 4.4 shows, $2\,^3S$ is the lowest triplet term in He, and all recaptures to triplets eventually cascade down to it. Depopulation occurs only by collisional transitions to $2\,^1S$ and $2\,^1P$, or by the strongly forbidden $2\,^3S$–$1\,^1S$ radiative transition, as discussed in Section 2.3, and as a result $N_{2\,^3S}$ is large, which in turn makes the optical depths in the lower $2\,^3S$–$n\,^3P$ lines significant. It can be seen from the energy-level diagram that $\lambda 10{,}830$ $2\,^3S$–$2\,^3P$ photons are simply scattered, but that absorption of $\lambda 3889$ $2\,^3S$–$3\,^3P$ photons can lead to their conversion to $\lambda 4.3\,\mu$ $3\,^3S$–$3\,^3P$, plus $2\,^3P$–$3\,^3S$ $\lambda 7065$, plus

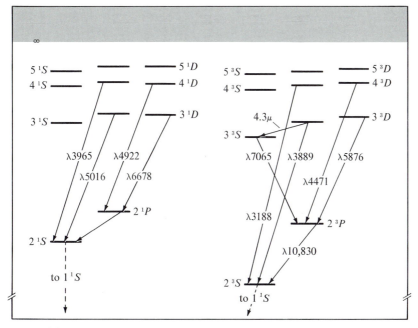

FIGURE 4.4
Partial energy-level diagram of He I, showing strongest optical lines observed in nebulae. Note that $1\,^2S$ has been omitted, and terms with $n \geqslant 6$ or $L \geqslant 2$ have been omitted for the sake of space and clarity.

$2\ ^3P-2\ ^3S\ \lambda 10{,}830$. The probability of this conversion is

$$\frac{A_{3\,^3S,\,3\,^3P}}{A_{3\,^3S,\,3\,^3P} + A_{2\,^3S,\,3\,^3P}} \approx 0.10$$

per absorption. At larger $\tau_{0l}(\lambda 10{,}830)$, still higher members of the $2\ ^3S-n\ ^3P$ series are converted to longer wavelength photons.

The radiative transfer problem is very similar to that for the Lyman lines discussed in Section 4.5, and may be handled by the same kind of formalism. Calculated ratios of the intensities of $\lambda 3889$ (which is weakened by self-absorption) and of $\lambda 7065$ (which is strengthened by resonance fluorescence) relative to the intensity of $\lambda 4471\ 2\ ^3P-4\ ^3D$ (which is only slightly affected by absorption) are shown for spherically symmetric homogeneous model nebulae in Figure 4.5.

The thermal Doppler widths of He I lines are smaller than those of H I lines, because of the larger mass of He, and therefore whatever turbulent

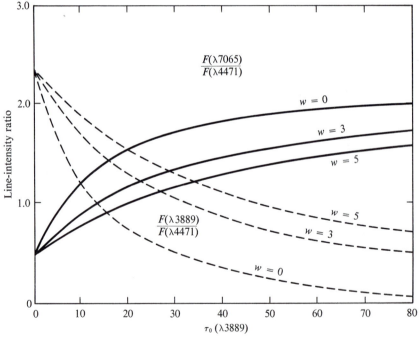

FIGURE 4.5
Radiative-transfer effects due to finite optical depths in He I $\lambda 3889\ 2\ ^3S-3\ ^3P$. Ratios of emergent fluxes of $\lambda 7065$ and $\lambda 3889$ to flux of $\lambda 4471$ shown as function of optical radius $\tau_0(\lambda 3889)$ of homogeneous static ($w = 0$) and expanding ($w \neq 0$) isothermal nebulae at $T = 10{,}000^\circ$K.

or expansion velocity there may be in a nebula is relatively more important in broadening the He I lines. The simplest case to consider is a model spherical nebula expanding with a linear velocity of expansion,

$$V_{\exp}(r) = wr \qquad 0 \leqslant r \leqslant R; \qquad (4.54)$$

for then between any two points r_1 and r_2 in the nebula, the relative radial velocity is

$$v_{\mathrm{rad}}(r_1, r_2) = ws, \qquad (4.55)$$

where s is the distance between the points and w is the constant velocity gradient. Thus photons emitted at r_1 will have a line profile centered about the line frequency v_L in the reference system in which r_1 is at rest. However, they will encounter at r_2 material absorbing with a profile centered on the frequency

$$v'(r_1, r_2) = v_L \left(1 + \frac{ws}{c}\right), \qquad (4.56)$$

and the optical depth in a particular direction to the boundary of the nebula for a photon emitted at r_1 with frequency v may be written

$$\tau_v(r_1) = \int_0^{r_2=R} N_{2\,3S} k_{0l} \exp\left\{-\left[\frac{v - v'(r_1, r_2)}{\Delta v_D}\right]^2\right\} ds. \qquad (4.57)$$

It can be seen that increasing velocity of expansion tends, for a fixed density $N_{2\,3S}$, to decrease the optical distance to the boundary of the nebula and thus to decrease the self-absorption effects. This effect can be seen in Figure 4.5, where some calculated results are shown for various ratios of the expansion velocity $V_{\exp}(R) = wR$ to the thermal velocity $V_{\mathrm{th}} = [2kT/M(\mathrm{He})]^{1/2}$, as functions of $\tau_{0l}(\lambda 3889) = N_{2\,3S} k_{0l}(\lambda 3889)R$, the optical radius at the center of the line for zero expansion velocity. Note that the calculated intensity ratios for large $V_{\exp}/V_{\mathrm{th}}$ and large τ_0 are quite similar to those for smaller $V_{\exp}/V_{\mathrm{th}}$ and smaller τ_0.

4.7   The Bowen Resonance-Fluorescence Mechanism for O III

There is an accidental coincidence between the wavelength of the He II $L\alpha$ line at $\lambda 303.78$ and the O III $2p^2\,{}^3P_2{-}3d\,{}^3P_2^0$ line at $\lambda 303.80$. As we have seen, in the He^{++} zone in a nebula there is some residual He$^+$, so the He I $L\alpha$ photons emitted by recombination are scattered many times before they escape. As a result, there is a high density of He II $L\alpha$ photons in the He^{++} zone, and since O^{++} is also present in this zone, some of the He II $L\alpha$ photons are absorbed by it and excite the $3d\,{}^3P^0_2$ level of O III. This level then quickly decays by a radiative transition, in most cases (relative probability 0.74) by resonance scattering in the $2p^2\,{}^3P_2$–$3d\,{}^3P^0_2$ line, that is, by emitting a photon. The next most likely decay process (probability 0.24) is emission of $\lambda 303.62$ $2p^2\,{}^3P_1$–$3d\,{}^3P^0_2$, which may then escape or may be reabsorbed by another O^{++} ion, again populating $3d\,{}^3P^0_2$. Finally (probability 0.02), the $3d\,{}^3P^0_2$ level may decay by omitting one of the six longer wavelength photons $3p\,{}^3L_J$– $3d\,{}^3P^0_2$ indicated in Figure 4.6 and listed in Table 4.13. These levels $3p\,{}^3L_J$ then decay to $3s$ and ultimately back to $2p^2\,{}^3P$, as shown in the figure and table. This is the Bowen resonance-fluorescence mechanism, the conversion of He II $L\alpha$ to those particular lines that arise from $3d\,{}^3P^0_2$ or from the levels excited by its decay. These lines are observed in many planetary nebulae, and their interpretation requires the solution of the problem of the scattering, escape, and destruction of He II, $L\alpha$, with the complications introduced by the O^{++} scattering and resonance fluorescence.

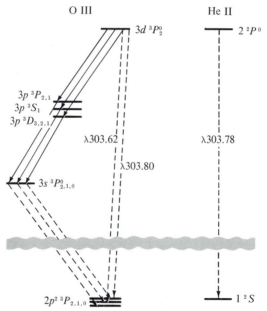

FIGURE 4.6
Schematized partial energy-level diagrams of [O III] and He II showing coincidence of He II $L\alpha$ and [O III] $2p^2\,{}^3P_2$–$3d\,{}^3P^0_2$ $\lambda 303.80$. The Bowen resonance-fluorescence lines in the optical and near-ultraviolet are indicated by solid lines, the far ultraviolet lines that lead to excitation or decay are indicated by dashed lines.

TABLE 4.13
O III *Resonance-Fluorescence Lines*

Transition	Wavelength (A)	Relative probability	Relative intensity
$3p\,^3P_2\text{--}3d\,^3P_2^0$	3444.10	5.7×10^{-3}	1.00
$3p\,^3P_1\text{--}3d\,^3P_2^0$	3428.67	1.9×10^{-3}	0.33
$3p\,^3S_1\text{--}3d\,^3P_2^0$	3132.86	1.3×10^{-2}	2.51
$3p\,^3D_3\text{--}3d\,^3P_2^0$	2837.17	7.6×10^{-4}	0.16
$3p\,^3D_2\text{--}3d\,^3P_2^0$	2819.57	1.4×10^{-4}	0.03
$3p\,^3D_1\text{--}3d\,^3P_2^0$	2808.77	8.0×10^{-5}	0.02
$3s\,^3P_2^0\text{--}3p\,^3S_1$	3340.74	7.1×10^{-3}	1.28
$3s\,^3P_1^0\text{--}3p\,^3S_1$	3312.30	4.4×10^{-3}	0.80
$3s\,^3P_0^0\text{--}3p\,^3S_1$	3299.36	1.5×10^{-3}	0.27
$3s\,^3P_2^0\text{--}3p\,^3P_2$	3047.13	4.2×10^{-3}	0.85
$3s\,^3P_1^0\text{--}3p\,^3P_2$	3023.45	1.4×10^{-3}	0.30
$3s\,^3P_2^0\text{--}3p\,^3P_1$	3059.30	7.8×10^{-4}	0.15
$3s\,^3P_1^0\text{--}3p\,^3P_1$	3035.43	4.8×10^{-4}	0.095
$3s\,^3P_0^0\text{--}3p\,^3P_1$	3024.57	6.4×10^{-4}	0.13
$3s\,^3P_0^0\text{--}3p\,^3D_1$	3757.21	4.5×10^{-5}	0.0073
$3s\,^3P_1^0\text{--}3p\,^3D_1$	3774.00	3.3×10^{-5}	0.0053
$3s\,^3P_2^0\text{--}3p\,^3D_1$	3810.96	2.1×10^{-6}	0.00036
$3s\,^3P_1^0\text{--}3p\,^3D_2$	3754.67	1.1×10^{-4}	0.017
$3s\,^3P_2^0\text{--}3p\,^3D_2$	3791.26	3.4×10^{-5}	0.0054
$3s\,^3P_2^0\text{--}3p\,^3D_3$	3759.87	8.1×10^{-4}	0.13

A competing process that can destroy He II $L\alpha$ photons before they are converted to O III Bowen resonance-fluorescence photons is absorption in the photoionization of H^0 and He^0. To take this process into account quantitatively, a specific model of the ionization structure is necessary, and the available calculations refer to two spherical-shell model planetary nebulae, ionized by stars with $T_* = 63,000°K$ and $100,000°K$, respectively. The optical depths in the center of the He II $L\alpha$ line are quite large in these models, $\tau_{0l}(He^+ \, L\alpha) \gtrsim 2 \times 10^5$, and indeed are large in all models in which the He^+-ionizing photons are completely absorbed, because the optical depth in the line center is much greater than in the continuum. Therefore, practically no He II $L\alpha$ photons escape directly at the outer surface of the nebula, and this result is quite insensitive to the exact value of $\tau_{0l}(\text{He II } L\alpha)$.

The detailed radiative-transfer solutions show that a little less than half the He II $L\alpha$ photons generated by recombination in the He^{++} zone are converted to Bowen resonance-fluorescence photons, while approximately one-third escape at the inner edge of the nebula, as shown numerically in Table 4.14. The photons that "escape" at the inner edge of the nebula at an angle θ to the normal simply cross the central hollow sphere of the model

TABLE 4.14
Probability of Escape or Absorption of He II *Lα*

Process	Planetary nebula model		
	$T_* = 6.3 \times 10^4$	$T_* = 1.0 \times 10^5$	
	$T = 1.0 \times 10^4$	$T = 1.0 \times 10^4$	$T = 2.0 \times 10^4$
Bowen conversion	0.49	0.42	0.43
He II $L\alpha$ escape	0.40	0.33	0.36
O III $^3P_1-^3P_2^0$ escape	0.09	0.08	0.12
H^0 or He^0 ionization	0.02	0.17	0.09

and enter the nebular shell again at another point at an angle θ to the normal there. However, taking account of the expansion of the nebula, these photons are redshifted by an amount $\Delta\nu = \nu_0(2 V_{exp}/c) \cos\theta$, where V_{exp} is the expansion velocity at the inner edge of the nebula. If, for instance, $V_{exp} \approx$ 10 km sec^{-1}, this can amount to a shift of several Doppler widths, since $V_{th} = 6.5$ km sec^{-1} at $T = 10,000°$K.

These photons that enter the nebula at the inner edge may be scattered or absorbed, and a certain fraction of them again escapes at the inner edge. All the photons that are absorbed within the Doppler core of the line ($|x| \lesssim 3$) are redistributed in frequency, while those absorbed in the wing ($|x| > 3$) are scattered more or less coherently. Eventually, some photons can be redshifted so much that, for them, the entire optical depth of the nebula is small and they then escape. The detailed calculation has been carried through for only one of the models of Table 4.14 (the model with $T_* = 100,000°$K, $T = 20,000°$K), and the result is that, of the 0.36 He II $L\alpha$ photons that escape at the inner edge of the model, a fraction 0.53 ultimately escape by redshifting, 0.20 are ultimately converted to O III $^3P_1-^3P_2^0$ photons, 0.22 are ultimately converted to Bowen resonance-fluorescence photons, and 0.05 are absorbed in H^0 or He^0 photoionization. The overall fraction of He II $L\alpha$ photons converted into Bowen resonance-fluorescence photons is thus 0.43 (directly) + 0.36 × 0.22 (following at least one escape into the central hole) = 0.51. These photons are distributed among the various individual lines as indicated in Table 4.13, in which the relative intensities are normalized to λ3444.

4.8 Collisional Excitation in He I

Collisional excitation of H is negligible in comparison with recombination in populating the excited levels in planetary nebulae and H II regions, because the threshold for even the lowest level, $n = 2$ at 10.2 eV, is large in comparison with the thermal energies at typical nebular temperatures.

This can be confirmed quantitatively using the excitation rates listed in Table 3.12. However, in He^0 the $2\,^3S$ level is highly metastable, and collisional excitation from it can be important, particularly in exciting $2\,^3P$ and thus leading to emission of He I $\lambda10,830$. To fix our ideas, let us consider a nebula sufficiently dense ($N_e \gg N_c$) so that the main mechanism for depopulating $2\,^3S$ is collisional transitions to $2\,^1S$ and $2\,^1P$, as explained in Section 2.4. The equilibrium population in $2\,^3S$ is then given by the balance between recombinations to all triplet levels, which eventually cascade down to $2\,^3S$, and collisional depopulation of $2\,^3S$,

$$N_e N_{\mathrm{He^+}}\alpha_B(He^0, n^3L) = N_e N_{2\,^3S}(q_{2\,^3S,\,2\,^1S} + q_{2\,^3S,\,2\,^1P}). \qquad (4.58)$$

The rate of collisional population of $2\,^3P$ is thus

$$N_e N_{2\,^3S}q_{2\,^3S,\,2\,^3P} = \frac{N_e N_{\mathrm{He^+}}q_{2\,^3S,\,2\,^3P}}{(q_{2\,^3S,\,2\,^1S} + q_{2\,^3S,\,2\,^1P})}\,\alpha_B(He^0, n\,^3L), \qquad (4.59)$$

so the relative importance of collisional to recombination excitation of $\lambda10,830$ is given by the ratio

$$\frac{N_e N_{2\,^3S}q_{2\,^3S,\,2\,^3P}}{N_e N_{\mathrm{He^+}}\alpha^{\mathrm{eff}}_{\lambda10,830}} = \frac{q_{2\,^3S,\,2\,^3P}}{(q_{2\,^3S,\,2\,^1S} + q_{2\,^3S,\,2\,^1P})}\,\frac{\alpha_B(He^0, n\,^3L)}{\alpha^{\mathrm{eff}}_{\lambda10,830}}. \qquad (4.60)$$

Computed values for $q_{2\,^3S,\,2\,^3P}$ are listed in Table 4.15; it can be seen that they are much larger than those for $q_{2\,^3S,\,2\,^1S}$ and $q_{2\,^3S,\,2\,^1P}$ (listed in Table 2.5) because the cross section for the strong allowed $2\,^3S$–$2\,^3P$ transition is much larger than the exchange cross sections to the singlet levels. At a representative temperature $T = 10,000°$K, the first factor in equation (4.60) has the numerical value 8.8; the second, 1.4; and the ratio of collisional to recombination excitation is thus about 12. In other words, collisional excitation from $2\,^3S$ completely dominates the emission of $\lambda10,830$, and the factor by which it dominates depends only weakly on T, and can easily be seen to decrease with N_e below N_c.

TABLE 4.15
Collisional Excitation Rate of $2\,^3P$

$T\,(°K)$	$q_{2\,^3S,\,2\,^3P}$ (cm^3 sec^{-1})
6000	1.1×10^{-7}
8000	2.2×10^{-7}
10,000	3.3×10^{-7}
12,000	4.4×10^{-7}
15,000	5.8×10^{-7}
20,000	7.9×10^{-7}

Though the collisional transition rates from $2\,^3S$ to $2\,^1S$ and $2\,^1P$ are smaller than to $2\,^3P$, the recombination rates of population of these singlet levels are also smaller, and the collisions are therefore also important in the population of $2\,^1S$ and $2\,^1P$. The cross sections for collisions to the higher singlets are negligibly small; accurate cross sections are not available for collisions to the higher triplets, but it appears likely that collisional population of $3\,^3P$ may be important and may somewhat affect the strength of $\lambda3889$. The cross sections from $2\,^3S$ to $n\,^3D$ are not accurately known; however, the available information seems to indicate that they are too small to affect significantly the strengths of $\lambda5876$ and $\lambda4471$.

Similar collisional excitation effects occur from the metastable $He^0\,2\,^1S$ and $H^0\,2\,^2S$ levels, but they decay so much more rapidly than $He^0\,2\,^3S$ that their populations are much smaller and the resulting excitation rates are negligibly small.

References

A good general summary of the emission processes in gaseous nebulae is given by
    Seaton, M. J. 1960. *Reports on Progress in Physics* **23**, 313, 1960.

The theory of the recombination-line spectrum of H I goes back to the early 1930's, and was developed in early papers by H. H. Plaskett, G. Cillie, D. H. Menzel, L. H. Aller, L. Goldberg, and others. In recent years it has been refined and worked out more accurately by M. J. Seaton, A. Burgess, R. M. Pengelly, M. Brocklehurst, and others. The treatment in Chapter 4 follows most closely the following definitive references:
    Seaton, M. J. 1959. *M. N. R. A. S.* **119**, 90.
    Pengelly, R. M. 1964. *M. N. R. A. S.* **127**, 145.
The second reference treats the low-density limit for H I and He II in detail. (Tables 4.1, 4.2, and 4.3 are derived from it.)

    Pengelly, R. M., and Seaton, M. J. 1964. *M. N. R. A. S.* **127**, 165.
The effects of collisions in shifting L at fixed n are discussed in this reference.

    Brocklehurst, M. 1971. *M. N. R. A. S.* **153**, 471.
This reference includes the definitive results for H I and He II at finite densities in the optical region, taking full account of the collisional transitions. (Tables 4.4 and 4.5 are based on it.)

Robbins, R. R. 1968. *Ap. J.* **151**, 497.

Robbins, R. R. 1970. *Ap. J.* **160**, 519.

Robbins, R. R., and Robinson, E. L. 1971. *Ap. J.* **167**, 249.

Brocklehurst, M. 1972. *M. N. R. A. S.* **157**, 211.

The first two papers by Robbins work out the theory in detail for the He I triplets; the third is concerned with the singlets (for case *A* only). The Brocklehurst article is complete, for it describes both triplet and singlet results. (Table 4.6 is based on it.)

Brown, R. L., and Mathews, W. G. 1970. *Ap. J.* **160**, 939.

This reference collects previous references and material on the H I and He II continuum, and includes the most detailed treatment of the He I continuum. (Tables 4.7–4.9 and 4.11 are taken from this reference.)

A very complete reference on free-free emission in the radio-frequency region is:

Scheuer, P. A. G. 1960. *M. N. R. A. S.* **120**, 231.

The first prediction that the radio-frequency recombination lines of H I would be observable is due to

Kardashev, N. S. 1959. *Astron. Zhurnal* **36**, 838 (English translation, *Soviet Astronomy A. J.* 1960. **3**, 813).

The key reference of the importance of the maser action and of the exact variation of b_n with n is:

Goldberg, L. 1966. *Ap. J.* **144**, 1225.

The radiative transfer treatment in this chapter essentially follows this reference.

The equilibrium equations for the populations of the high levels are worked out in

Seaton, M. J. 1964. *M. N. R. A. S.* **127**, 177.

Sejnowski, T. J., and Hjellming, R. H. 1969. *Ap. J.* **156**, 915.

Brocklehurst, M. 1970, *M. N. R. A. S.* **148**, 417.

The last of these three references is the definitive treatment and makes full allowance for all collisional effects. (Figure 4.1 is based on it.)

A good deal of theoretical work has been done by several authors on radiative-transfer problems in nebulae. This research is summarized (with complete references) in

Osterbrock, D. E. 1971. *J. Q. S. R. T.* **11**, 623.

Some of the key references concerning the H I lines are:

Capriotti, E. R. 1964. *Ap. J.* **139**, 225.

Capriotti, E. R. 1964. *Ap. J.* **140**, 632.

Capriotti, E. R. 1966. *Ap. J.* **146**, 709.

Cox, D. P., and Mathews, W. G. 1969. *Ap. J.* **155**, 859.

(Figure 4.3 is based on the last reference.)

Actually, the radiative transfer problem of He I lines was worked out earlier in
  Pottasch, S. R. 1962. *Ap. J.*, **135**, 385.
A more recent and complete treatment is:
  Robbins, R. R. 1968. *Ap. J.* **151**, 511.
(Figure 4.5 is derived from calculations in this reference.)

The Bowen resonance-fluorescence mechanism was first described by
  Bowen, I. S. 1924. *Ap. J.* **67**, 1.
The most complete radiative-transfer theory of this problem is due to
  Weymann, R. J., and Williams, R. E. 1969. *Ap. J.* **157**, 1201.
The calculations described in Section 4.7 are contained in the preceding reference.
(Table 4.14 is derived from it.) Still further details of the Bowen resonance-
fluorescence process are included in
  Harrington, J. P. 1972. *Ap. J.* **176**, 127.

The importance of collisional excitation from $He^0\, 2\,^3S$ was discussed by
  Mathis, J. S. 1957. *Ap. J.* **125**, 318.
  Pottasch, S. R. 1961. *Ap. J.* **135**, 93.
  Osterbrock, D. E. 1964. *Ann. Rev. Astr. and Astrophys.* **2**, 95.
  Cox, D. P., and Daltabuit, E. 1971. *Ap. J.* **167**, 257.
See also the references by Robbins, and especially by Brocklehurst, in this section.

The cross sections used in Table 4.15 are from
  Burke, P. G., Cooper, J. W., and Ormonde, S. 1969. *Phys. Rev.* **183**, 245.

5

Comparison of Theory with Observations

5.1   Introduction

In the preceding three chapters much of the available theory on gaseous nebulae has been discussed, so that we are now in a position to compare this theory with the available observations. The temperature in a nebula may be determined from measurements of ratios of intensities of particular pairs of emission lines—those emitted by a single ion from two levels with considerably different excitation energies. Although the relative strengths of H recombination lines vary only extremely weakly with T, the ratio of the intensity of a line to the intensity of the recombination continuum varies more rapidly and can be used to measure T. Further information on the temperature may be derived from radio observations, combining long and short wavelength continuum observations (large and small optical depths, respectively) or long wavelength continuum and optical line observations. The electron density in a nebula may be determined from measured intensity ratios of other pairs of lines—those emitted by a single ion from two levels with nearly the same energy but with different radiative-transition proba-bilities. Likewise, measurements of relative strengths of the radio recombination lines give information on both the density and the temperature in nebulae. These methods, as well as the resulting information on the physical parameters of characteristic nebulae, are discussed in the first sections of this chapter.

In addition, information on the involved stars that provide the ionizing photons may be derived from nebular observations. For if a nebula is optically thick to a particular type of ionizing radiation (for instance, in the H Lyman continuum), then the total number of photons of this type emitted by the star can be determined from the properties of the nebula. By combining these nebular observations, which basically measure the far-ultra-violet-ionizing radiation from the involved stars, with optical measurements of the same stars, a long base-line color index that gives information on the temperature of the stars can be determined. This scheme and the information derived from it, about main sequence O stars and about plane-tary-nebula central stars, are discussed in Section 5.7.

Once the temperature and density in a nebula are known, it is fairly clear that the observed strength of a line gives information on the total number of ions in the nebula responsible for the emission of that line. Thus information is derived on the abundances of the elements in H II regions and planetary nebulae.

Each of the first eight sections of this chapter discusses a particular kind of observational analysis or diagnostic measurement of a nebula. Each method gives some specific detailed information, integrated through whatever structure there may be, along the line of sight through the nebula, and also over whatever area of the nebula is covered by the analyzing device used for the observations, such as the spectrograph slit or the radio-telescope beam pattern. A more detailed comparison, in integrated form, may be made by calculating models of nebulae intended to represent their entire structure and comparing the properties of these models with observations. A discussion of this type of models, and of the progress that has been made with them, closes the chapter.

5.2   Temperature Measurements from Emission Lines

A few ions, of which [O III] and [N II] are the best examples, have energy-level structures that result in emission lines from two different upper levels with considerably different excitation energies occurring in the observable wavelength region. The energy-level diagrams of these two ions are shown in Figure 3.1, where it can be seen that, for instance, [O III] $\lambda4363$ occurs from the upper ^1S level, while $\lambda4959$ and $\lambda5007$ occur from the lower ^1D level. ($^3P_0-^1D_2$ $\lambda4931$, which can occur only by an electric-quadrupole transition, has much smaller transition probability and is so weak that it can be ignored.) It is clear that the relative rates of excitations to the ^1S and ^1D levels depend very strongly on T, so the relative strength of the lines emitted by these levels may be used to measure electron temperature.

An exact solution for the populations of the various levels, and for the relative strengths of the lines emitted by them, may be carried out along the lines of the discussion in Section 3.5. However, it is simpler and more instructive to proceed by direct physical reasoning. In the low-density limit (collisional de-excitations negligible), every excitation to the ^1D level results in emission of a photon either in $\lambda5007$ or $\lambda4959$, with relative probabilities given by the ratio of the two transition probabilities, which is very close to 3 to 1. Every excitation of ^1S is followed by emission of a photon in either $\lambda4363$ or $\lambda2321$, with the relative probabilities again given by the transition probabilities. Each emission of a $\lambda4363$ photon further results in the population of ^1D, which again is followed by emission of either a $\lambda4959$ photon or a $\lambda5007$ photon; but this contribution is small in comparison with the direct excitation of ^1D and can be neglected. Thus the ratio of emission-line strengths in the low-density limit is given simply by

$$\frac{j_{\lambda4959} + j_{\lambda5007}}{j_{\lambda4363}} = \frac{\Omega(^3P, {}^1D)}{\Omega(^3P, {}^1S)} \frac{A_{{}^1S, {}^1D} + A_{{}^1S, {}^3P}}{A_{{}^1S, {}^1D}} \frac{\overline{\nu}(^3P, {}^1D)}{\nu({}^1S, {}^1D)} e^{\Delta E/kT}, \quad (5.1)$$

where

$$\overline{\nu}(^3P, {}^1D) = \frac{A_{{}^3P_2, {}^1D_2}\nu(\lambda5007) + A_{{}^3P_1, {}^1D_2}\nu(\lambda4959)}{A_{{}^3P_2, {}^1D_2} + A_{{}^3P_1, {}^1D_2}}, \quad (5.2)$$

and ΔE is the energy difference between the $^1D_2$ and $^1S_0$ levels.

Equation (5.1) is a good approximation up to $N_e \approx 10^5 \text{ cm}^{-3}$. However, at higher densities collisional de-excitation begins to play a role. The lower ^1D term has a considerably longer radiative lifetime than the ^1S term, so it is collisionally de-excited at lower electron densities than ^1S, thus weakening $\lambda4959$ and $\lambda5007$. In addition, under these conditions collisional excitation of ^1S from the excited ^1D level begins to strengthen $\lambda4363$. The full statistical equilibrium equations (3.28) can be worked out numerically for any N_e and T, but an analytic solution correct to the first order in N_e and to the first order in $e^{-\Delta E/kT}$ is that the right-hand side of (5.1) is multiplied by a factor

$$f = \frac{1 + \dfrac{C(^3P, {}^1D)C({}^1D, {}^1S)}{C(^3P, {}^1S)A_{{}^1D, {}^3P}} + \dfrac{C({}^1D, {}^3P)}{A_{{}^1D, {}^3P}}}{1 + \dfrac{C(^3P, {}^1S) + C({}^1S, {}^1D)}{A_{{}^1S, {}^3P} + A_{{}^1D, {}^3P}}}, \quad (5.3)$$

where

$$C(i,j) = 8.63 \times 10^{-6} \frac{N_e}{T^{1/2}} \frac{\Omega(i,j)}{\omega_i}.$$

Inserting numerical values of the collision strengths and transition probabilities from Chapter 3, this becomes

$$\frac{j_{\lambda4959} + j_{\lambda5007}}{j_{\lambda4363}} = \frac{8.32 \exp\left(\dfrac{3.29 \times 10^4}{T}\right)}{1 + 4.5 \times 10^{-4} \dfrac{N_e}{T^{1/2}}}. \tag{5.4}$$

Here the representative values of the collision strengths from Table 3.4 have been used to calculate the numerical coefficients, but actually in O^{++}, there are several large resonances and the resulting average collision strengths vary appreciably with temperature, so equation (5.4) is not exact. However, in Figure 5.1 the intensity ratio is plotted (in the low-density limit)

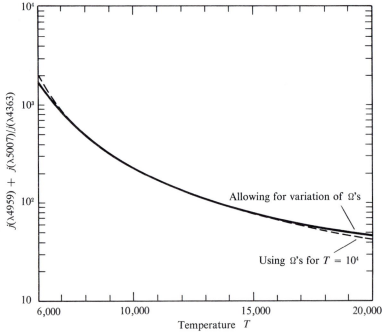

FIGURE 5.1
[O III] ($\lambda4959 + \lambda5007$)/$\lambda4363$ intensity ratio (in low-density limit $N_e \rightarrow 0$) as a function of temperature. Solid line shows accurately calculated value; dashed line shows approximation of equation (5.4) using mean values of Ω.

using the correct collision strengths at each T, and also according to equation (5.4), and it can be seen that very little error results from the use of mean collision strengths.

An exactly similar treatment may be carried out for [N II], and the resulting equation analogous to (5.4) is

$$\frac{j_{\lambda 6548} + j_{\lambda 6583}}{j_{\lambda 5755}} = \frac{7.53 \exp\left(\dfrac{2.50 \times 10^4}{T}\right)}{1 + 2.7 \times 10^{-3}\dfrac{N_e}{T^{1/2}}}. \tag{5.5}$$

These two equations form the basis for optical temperature determinations in gaseous nebulae. Since the nebulae are optically thin in forbidden-line radiation, the ratio of the integrals of the emission coefficients along a ray through the nebula is observed directly as the ratio of emergent intensities, so if the nebula is assumed to be isothermal and to have sufficiently low density that the low-density limit is applicable, the temperature is directly determined. Alternatively, the ratio of the fluxes from the whole nebula may be measured in the case of smaller nebulae. No information need be known on the distance of the nebula, and amount of O^{++} present, and so on, as all these factors cancel out. If collisional de-excitation is not completely negligible, even a rough estimate of the electron density substituted into the correction term in the denominator provides a good value of T. The observed strengths of the lines must be corrected for interstellar extinction, but this correction is usually not too large because the temperature-sensitive lines in both [O III] and [N II] are relatively close in wavelength.

The [O III] line-intensity ratio $(\lambda 4959 + \lambda 5007)/\lambda 4363$ is quite large and is therefore rather difficult to measure accurately. Although $\lambda\lambda 4959, 5007$ are strong lines in many gaseous nebulae, $\lambda 4363$ is relatively weak, and furthermore is unfortunately close to Hg I $\lambda 4358$, which is becoming stronger and stronger in the spectrum of the sky. Large intensity ratios are difficult to measure photographically with any accuracy at all, and therefore reasonably precise temperature measurements require carefully calibrated photoelectric measurements made with fairly high-resolution spectral analyzers. Up to the present time, most work has been done on the [O III] lines, partly because they occur in the blue spectral region in which photomultipliers are most sensitive, and partly because [O III] is quite bright in typical high-surface brightness planetary nebulae. The [N II] lines are stronger in the outer parts of H II regions, where the ionization is lower and the O is mostly [O II], but not too many measurements are yet available.

Let us first examine optical determinations of the temperatures in H II regions, some selected results of which are collected in Table 5.1. Note that in this and other tables, the observed intensity ratio has been corrected for interstellar extinction in the way outlined in Chapter 7, and the temperature has been computed using numerical values from this book—in the present case Figure 5.1 and equation (5.5).

It can be seen that all the temperatures of these H II regions are in the range 8000–9000°K. Of the three slit position in NGC 1976, the Orion Nebula, II is in the brightest part near the Trapezium, and I and III are also in bright regions not far from the center. In fact, all the observations in Table 5.1 refer to selected relatively bright, relatively dense parts of H II regions, and more observations on fainter objects would certainly be valuable.

Planetary nebulae have higher surface brightness than typical H II regions, and as a result there is a good deal more observational material available for planetaries, particularly [O III] determinations of the temperature. Most planetaries are so highly ionized that [N II] is relatively weak, but some measurements of it are also available. The best observational material is collected in Table 5.2, which shows that the temperatures in planetary nebulae are typically somewhat higher than in H II regions. This is partly a consequence of higher effective temperatures of the central stars in planetary nebulae (to be discussed in Section 5.7), leading to a higher input of energy per photoionization, and partly a consequence of the higher electron densities in typical planetaries, resulting in collisional de-excitation and decreased efficiency of radiative cooling. Discrepancies between T as determined from [O III] and [N II] lines in the same nebula do not necessarily indicate an error in either method; these lines are emitted in different zones of the nebula because their ionization potentials are different.

From Tables 5.1 and 5.2, it is reasonable to adopt $T \approx 10,000°$K as an order-of-magnitude estimate for any nebula; with somewhat greater precision we may adopt representative values $T \approx 9000°$K in the brighter parts of an H II region like NGC 1976, and $T \approx 11,000°$K in a typical bright planetary nebula.

TABLE 5.1
Temperature Determinations in H II Regions

Nebula	[N II]			[O III]	
			$N_e/T^{1/2}$		
	$\dfrac{I(\lambda6548) + I(\lambda6583)}{I(\lambda5755)}$	T		$\dfrac{I(\lambda4959) + I(\lambda5007)}{I(\lambda4363)}$	T
NGC 1976 I	68	11,100°	24	406	8200°
NGC 1976 II	81	10,600°	45	331	8900°
NGC 1976 III	123	8800°	18	323	9000°
M 8 I	162	8100°	(10)	445	8300°
M 17 I	257	7000°	(10)	330	9000°

TABLE 5.2
Temperature Determinations in Planetary Nebulae

Nebula	[N II]			[O III]	
			$N_e/T^{1/2}$		
	$\dfrac{I(\lambda 6548) + I(\lambda 6583)}{I(\lambda 5755)}$	T		$\dfrac{I(\lambda 4959) + I(\lambda 5007)}{I(\lambda 4363)}$	T
NGC 1535	—	—	35	132	12,000
NGC 2392	—	—	65	49	18,400
NGC 6572	38	15,500	73	186	10,600
NGC 6720	—	—	5.8	174	10,900
NGC 6803	80	10,300	36	240	9800
NGC 6826	—	—	(50)	120	12,400
NGC 7009	—	—	52	214	10,200
NGC 7027	60	11,800	60	74	15,300
NGC 7662	—	—	37	96	13,500
IC 418	85	10,000	400	—	—
IC 3568	—	—	(50)	151	11,400
IC 4593	—	—	22	316	9100
IC 4997	—	—	large	22	see text
IC 5217	—	—	(50)	141	11,700

5.3 Temperature Determinations from Optical Continuum Measurements

Although it might be thought that the temperature in a nebula could be measured from the relative strengths of the H lines, in fact their relative strengths are almost independent of temperature, as Table 4.4 shows. The physical reason for this behavior is that all the recombination cross sections to the various levels of H have approximately the same velocity dependence, so the relative numbers of atoms formed by captures to each level are nearly independent of T, and since the cascade matrices depend only on transition probabilities, the relative strengths of the lines emitted are also nearly independent of T. These calculated relative line strengths are in good agreement with observational measurements, as will be discussed in detail in Section 7.2.

However, the temperature in a nebula can be determined by measuring the relative strength of the recombination continuum with respect to a recombination line. Physically, the reason this ratio does depend on the temperature is that the emission in the continuum (per unit frequency interval) depends on the width of the free-electron velocity-distribution function, that is, on T.

The theory is straightforward, for we may simply use Table 4.4 to calculate the H-line emission, and Tables 4.7–4.9 and 4.12 to calculate the continuum emission, and thus find their ratio as a function of T. Table 5.3 lists the calculated ratios for two choices of the continuum, first at Hβ λ4861, which includes the H I recombination and two-photon continua as well as the He I recombination continuum. (A nebula with $N_{He^+} = 0.10\ N_{H^+}$ and $N_{He^{++}} = 0$ has been assumed, but any other abundances or ionization conditions determined from line observations of the nebula could be used.) The second choice is the Balmer continuum, $j_\nu(\lambda3646-) - j_\nu(\lambda3646+)$, which eliminates everything except the H I recombination continuum due to captures into $n = 2$. (The He II recombination, of course, would also contribute if $N_{He^{++}} \neq 0$.) Note that the λ4861 continuum has been calculated in the limit $N_p \to 0$ (no collisional de-excitation of H I 2 ^2S and hence maximum relative strength of the H I two-photon continuum) and also for the case $N_p = 10^4$ cm^{-3}, $N_{He^+} = 10^3$ cm^{-3}, taking account of collisional de-excitation, while the Balmer continuum results are independent of density and N_{He^+}.

The continuum at λ4861 is made up chiefly of the H I Paschen and higher-series continua, whose sum increases slowly with T, and the two-photon continuum, whose strength decreases slowly with T; the sum hence is roughly independent of T, and the ratio of this continuum to Hβ therefore increases with T. On the other hand, the strength of the Balmer continuum at the series limit decreases approximately as $T^{-3/2}$, and its ratio to Hβ therefore decreases slowly with T, as Table 5.3 shows.

The observations of the continuum are difficult because it is weak and

TABLE 5.3
Ratio* of Continuum to Line Emission

		T		
	5000°	10,000°	15,000°	20,000°
$\lim N_p \to 0$, $N_{He^+} = 0.10\ N_p$				
$\dfrac{j_\nu(\lambda4861)}{j_{H\beta}}$	3.45×10^{-15}	5.82×10^{-15}	7.31×10^{-15}	9.21×10^{-15}
$N_p = 10^4$ cm^{-3}, $N_{He^+} = 10^3$ cm^{-3}, $N_e = 1.1 \times 10^4$ cm^{-3}				
$\dfrac{j_\nu(\lambda4861)}{j_{H\beta}}$	2.44×10^{-15}	4.81×10^{-15}	6.35×10^{-15}	8.20×10^{-15}
N_p, N_{He^+}, N_e arbitrary				
$\dfrac{j_\nu(\text{Bac})}{j_{H\beta}}$	3.10×10^{-14}	1.89×10^{-14}	1.41×10^{-14}	1.25×10^{-14}

* All ratios are in units of Hz^{-1}.

TABLE 5.4
Observations of Balmer Discontinuity in Nebulae

Nebula	$\dfrac{I_\nu(\lambda3646-) - I_\nu(\lambda3646+)}{I(H\beta)}$ (Hz^{-1})	T ($^\circ$K)
NGC 1976 I	2.32×10^{-14}	7300
NGC 1976 II	2.47×10^{-14}	6800
NGC 1976 III	2.56×10^{-14}	6500
NGC 6572 A	2.06×10^{-14}	8700
NGC 6572 B	2.45×10^{-14}	6900
NGC 7009 B	2.27×10^{-14}	7700
IC 418 A	2.26×10^{-14}	7700

NOTE: In NGC 1976 I, II, and III are slit positions that correspond to Table 5.1. In the planetary nebulae, A represents an entrance diaphragm 30″ in diameter, while B represents an entrance slit 8″ × 80″ oriented east-west.

can be seriously affected by weak lines. High-resolution spectrophotometric measurements with high-sensitivity detectors are necessary. To date the most accurate published data seem to be measurements of the Balmer continuum, which is considerably stronger than the continuum near Hβ. A difficulty in measuring the Balmer continuum, of course, is that the higher Balmer lines are crowded just below the limit, so the intensity must be measured at longer wavelengths and extrapolated to $\lambda3646+$. Furthermore, continuous radiation emitted by the stars involved in the nebulae and scattered by interstellar dust may have a sizable Balmer discontinuity, which is difficult to disentangle from the true nebular recombination Balmer discontinuity. Some of the best published results for H II regions and planetary nebulae are collected in Table 5.4, which shows that the temperatures measured by this method are generally somewhat smaller than the temperatures for the same objects measured from forbidden-line ratios. The most accurately observed nebula of all is NGC 7027, in which the observed continuum over the range $\lambda\lambda3300-11{,}000$, including both the Balmer and Paschen discontinuities, as well as the Balmer-line and Paschen-line ratios, matches very well the calculated spectrum for $T = 17{,}000^\circ$K, somewhat higher than the temperatures indicated by forbidden-line ratios. These discrepancies will be discussed again in the next section following the discussion of temperature measurements from the radio-continuum observations.

5.4 Temperature Determinations from Radio-Continuum Measurements

Another completely independent temperature determination can be made from radio-continuum observations. The idea is quite straightforward, namely, that at sufficiently low frequencies any nebula becomes optically thick, and therefore, at these frequencies the emergent intensity is the same as that from a black body, the Planck function $B_\nu(T)$; or equivalently, the measured brightness temperature is the temperature within the nebula

$$T_{b\nu} = T(1 - e^{-\tau_\nu}) \to T \text{ as } \tau_\nu \to \infty \qquad (5.6)$$

as in equation (4.37). Note that if there is background radiation (beyond the nebula) with brightness temperature $T_{bg\nu}$ and foreground radiation (between the nebula and the observer) with brightness temperature $T_{fg\nu}$, this equation becomes

$$T_{b\nu} = T_{fg\nu} + T(1 - e^{-\tau_\nu}) + T_{bg\nu}e^{-\tau_\nu} \to T_{fg\nu} + T \qquad \text{as } \tau_\nu \to \infty. \quad (5.7)$$

The difficulty with applying this method is that at frequencies that are sufficiently low that the nebulae are optically thick ($\nu \approx 3 \times 10^8$ Hz or $\lambda \approx 10^2$ cm for many dense nebulae), even the largest radio telescopes have beam sizes that are comparable to or larger than the angular diameters of typical H II regions. Therefore, the nebula does not completely fill the beam, and a correction must be made for the projection of the nebula onto the antenna pattern.

The antenna pattern of a simple parabolic or spherical dish is circularly symmetric about the axis, where the sensitivity is at a maximum. The sensitivity decreases outward in all directions, and in any plane through the axis it has a form much like a Gaussian function with angular width of order λ/d, where d is the diameter of the telescope. The product of the antenna pattern with the brightness-temperature distribution of the nebula then gives the mean brightness temperature, which is measured by the radio-frequency observations. The antenna pattern thus tends to broaden the nebula and to wipe out much of its fine structure. To determine the temperature of a nebula that is small compared with the width of the antenna pattern, it is therefore necessary to know its angular size accurately. But of course, no nebula really has sharp outer edges, inside which it has infinite optical depth and outside which it has zero optical depth. In a real nebula, the optical depth decreases more or less continuously but with many fluctuations, from a maximum value somewhere near the center of the nebula to zero just outside the edge of the nebula, and what is really needed is the complete distribution of optical depth over the face of the nebula.

This can be obtained from high radio-frequency measurements of the nebula, for in the high-frequency region, the nebula is optically thin, and the measured brightness temperature gives the product $T\tau_1$,

$$T_{b1} = T(1 - e^{-\tau_1}) \to T\tau_1 \qquad \text{as } \tau_1 \to 0, \qquad (5.8)$$

as in equation (4.37). (In the remainder of this section, the subscript 1 is used to indicate a high frequency, and subscript 2 is used to indicate a low frequency.) At high frequencies, the largest radio telescopes have considerably better angular resolution than at low frequencies because of the smaller values of λ/d, so that if the nebula is assumed to be isothermal, the high-frequency measurements can be used to prepare a map of the nebula, giving the product $T\tau_1$ at each point. Thus for any assumed T, the optical depth τ_1 is determined at each point from the high-frequency measurements. The ratio of optical depths, τ_1/τ_2, is known from equation (4.31), so τ_2 can be calculated at each point, and then the expected brightness temperature T_{b2} can be calculated at each point:

$$T_{b2} = T(1 - e^{-\tau_2}). \tag{5.9}$$

Integrating the product of this quantity with the antenna pattern gives the expected mean brightness temperature at the low frequency. If the assumed T is not correct, this expected result will not agree with the observed mean brightness temperature, and another assumed temperature must be tried until agreement is reached. This, then, is a procedure for correcting the radio-frequency continuum measurements for the effects of finite beam size at low frequencies. It would be exact if the high-frequency measurements had perfect angular resolution, but they also have finite resolution, of order 2′, even for the highest frequencies ($\lambda \approx 2$ cm) and for the largest telescopes ($d \approx 140$ feet) now in use. The high-frequency measurements themselves must therefore be corrected for the effects of finite resolution, because if they are not, they incorrectly indicate finite-brightness temperatures at regions outside the nebula where the optical depth is actually zero, and thus indicate an artificially large apparent size of the nebula, resulting in an incorrectly low calculated temperature from the low-frequency measurements of the nebula. Many of the published temperatures for H II regions determined from continuum measurements are too small because this correction was not fully taken into account.

Some of the most accurate available radio-frequency measurements of temperature in H II regions are collected in Table 5.5. At 408 MHz, most of the nebulae listed have central optical depths $\tau_2 \approx 1$ to 10, while at 85 MHz, the optical depths are considerably larger. The beam size of the antenna is larger at 85 MHz ($\sim 50'$ with the original Mills Cross), and in addition the background (nonthermal) radiation is larger, so many of the nebulae are measured in absorption at this lower frequency. The uncertainties are probably of order $\pm 1000°$.

The two frequencies give fairly consistent temperature determinations, and comparison of Table 5.5 with Table 5.1 shows that fairly good agreement exists between the two quite independent ways of measuring the mean temperature in a nebula. There are, however, some slight remaining differ-

TABLE 5.5
Electron Temperatures in H II *Regions from Radio-Frequency Continuum Observations*

Nebula	T	
	408 MHz	85 MHz
NGC 1976	8550°	7000°
RCW 38	7500°	4000°
RCW 49	7750°	6000°
NGC 6334	7000°	10,000°
NGC 6357	6900°	10,000°
M 17	7850°	8000°
M 16	—	5000°
NGC 6604	—	4000°

ences, generally in the sense that the temperature as determined from the radio-frequency observations is lower than the temperature determined from forbidden-line ratios. The probable explanation of this discrepancy is that the temperature is not constant throughout the nebula as has been previously assumed, but rather varies from point to point due to variations in the local heating and cooling rates. If this interpretation is correct, a more complicated comparison between observation and theory is necessary. An ideal method would be to know the entire temperature structure of the nebula, to calculate from it the expected forbidden-line ratios and radio-frequency continuum brightness temperatures, and then to compare them with observation; this is a model approach that will be discussed in Section 5.9.

However, the general type of effects that are expected can easily be understood. The forbidden-line ratios determine the temperature in the region in which these lines themselves are emitted—that is, the [O III] ratio measures a mean temperature in the O^{++} zone and the [N II] ratio measures the mean temperature in the N^+ zone, while the radio-frequency continuum measures the mean temperature in the entire ionized region. Furthermore, each method measures a mean temperature weighted in a different way. The emission coefficient for the forbidden lines increases strongly with increasing temperature, and therefore the mean they measure is strongly weighted toward high-temperature regions. On the other hand, the free-free emission coefficient decreases with increasing temperature, and therefore the mean it measures is weighted toward low-temperature regions. We thus expect a discrepancy in the sense that the forbidden lines indicate a higher temperature than do the radio-frequency measurements, as is in fact confirmed by observation. It is even possible to get some information about the range in variation of the temperature along a line through the nebula from comparison of these various temperatures, but as the result depends on the ionization distribution also, we shall not consider this method in detail.

Exactly the same method can be used to measure the temperatures in planetary nebulae, but as the nebulae are very small in comparison with the antenna beam size at the frequencies at which they are optically thick, the correction for this effect is quite important. Nearly all the planetary nebulae are too small for mapping at even the shortest radio-frequency wavelengths, but it is possible to use the surface brightness in a hydrogen recombination line such as Hβ, since it is also proportional to the integral

$$I(\mathrm{H}\beta) \propto \int N_p N_e \, ds = E_p,$$

to get the relative values of τ_2 at each point in the nebula. Even the optical measurements have finite angular resolution because of the broadening effects of seeing. Then, for any assumed optical depth of the nebula at one point and at one frequency, the optical depths at all other points and at all frequencies can be calculated. For any assumed temperature, the expected flux at each frequency can thus be calculated and compared with the radio measurements, which must be available for at least two (and preferably more) frequencies, one in the optically thin region and one in the optically thick region. The two parameters T and the central optical depth must be varied to get the best fit between calculations and measurements. The best observed planetary nebula is IC 418, for which the measured 408 MHz flux, combined with corrected Hβ isophotes, give $T = 12,500°$K. However, other published results for this nebula and for one or two other planetaries indicate lower temperatures. The uncertainties are largely due to the lack of accurate optical isophotes, from which the distribution of brightness temperature over the nebula must be calculated. It is somewhat disheartening to note that radio-frequency measurements for several planetaries are available, which could be interpreted to find their temperatures if only accurate optical Hβ isophotes for them were available. This is a challenge to the optical observer that should be met in the near future by using narrow-band filters on long-focal-length telescopes in conditions of good seeing.

A related method of using radio-frequency measurements alone to determine the temperature of a nebula is to adopt a standard spherically symmetric model, from which the entire predicted continuous spectrum can be calculated as a function of T and central optical depth τ_2 at a particular frequency, for a nebula of "known" angular size. The best fit of this predicted radio-frequency continuum to the observed continuum measurements determines both the product $T\tau_2$ (from the optically thin high-frequency region) and T (if there are measured fluxes at frequencies sufficiently low for the nebula to be optically thick). A recent very complete compilation of all measurements of continuum radio-frequency fluxes of planetary nebulae, together with estimates of probable errors of these measurements, indicates that for most of the planetaries the temperatures found in this way agree with the forbidden-line determinations of temperatures.

Finally, there is published radio-frequency observational evidence for a small "hot spot" with very high temperature in at least one nebula. The observation was made with a very long-base-line interferometer with high angular resolution, of order 0''.6, for the observations to be described. With such an interferometer it is possible, in principle, to get a complete map of an object by combining many days of observations taken with all possible spacings and orientations of the elements of the interferometer. However, for the measurements to be described here, an observation was made at only one spacing, and the result is not a complete map of the object, but rather some limited information on the apparent structure of the radio source. These measurements of the planetary nebula NGC 7027 at $\lambda = 11$ cm appear to show a small core with angular size $\leq 0''.6$ and brightness temperature $T_{bv} \gtrsim 7 \times 10^5 °$K. It is clear from equation (5.6) that if the radiation is thermal (and not, for instance, synchrotron radiation), the observation implies that there is a point in the nebula at which $T \gtrsim 700{,}000 °$K, a quite unexpected result in terms of the heating and cooling processes discussed in Chapter 3. Since the observation was made at only one frequency, it is possible that some other interpretation applies—for instance, that there is some sort of nonthermal source in the nebula. The result is of great importance, however, and should be actively pursued by further observations at other frequencies, at other spacings, and with other interferometers.

In fact, a later series of measurements of NGC 7027 at $\lambda = 6$ cm and at many spacings, which yields a complete map with a resolution of about $2'' \times 3''$, shows that at this frequency, the nebula matches very well the brightness temperature distribution expected from a toroidal cylinder with $T \approx 13{,}000 °$K. The hot spot does not appear in these or in other lower angular resolution measurements, but further very large-separation interferometer measurements are clearly required.

5.5   Electron Densities from Emission Lines

The electron density in a nebula may be measured by observing the effects of collisional de-excitation. This can be done by comparing the intensities of two lines of the same ion, emitted by different levels with nearly the same excitation energy so that the relative excitation rates to the two levels depend only on the ratio of collision strengths. If the two levels have different radiative transition probabilities or different collisional de-excitation rates, the relative populations of the two levels will depend on the density, and the ratio of intensities of the lines they emit will likewise depend on the density. The best examples of lines that may be used to measure the electron density are [O II] $\lambda 3729/\lambda 3726$ and [S II] $\lambda 6716/\lambda 6731$, with energy-level diagrams shown in Figure 5.2.

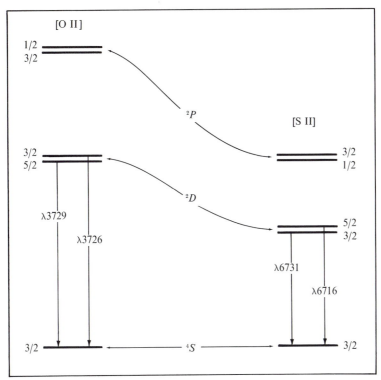

FIGURE 5.2
Energy-level diagrams of the $2p^3$ ground configuration of [O II] and $3p^3$ ground configuration of [S II].

The relative populations of the various levels and the resulting relative line-emission coefficients may be found by setting up the equilibrium equations for the populations of each level as described in Section 3.5. However, direct physical reasoning easily shows the effects involved. Consider the example of [O II] in the low-density limit $N_e \to 0$, in which every collisional excitation is followed by emission of a photon. Since the relative excitation rates of the $^2D_{5/2}$ and $^2D_{3/2}$ levels are proportional to their statistical weights (see equation 3.22), the ratio of strengths of the two lines is simply $j_{\lambda3729}/j_{\lambda3726} = 1.5$. On the other hand, in the high-density limit, $N_e \to \infty$, collisional excitations and de-excitations dominate and set up a Boltzmann population ratio. Thus the relative populations of the two levels $^2D_{5/2}$ and $^2D_{3/2}$ are in the ratio of their statistical weights, and therefore the relative strengths of the two lines are in the ratio

$$\frac{j_{\lambda3729}}{j_{\lambda3726}} = \frac{N_{^2D_{5/2}}}{N_{^2D_{3/2}}} \frac{A_{\lambda3729}}{A_{\lambda3726}} = \frac{3}{2} \frac{4.2 \times 10^{-5}}{1.8 \times 10^{-4}} = 0.35.$$

The transition between the high- and low-density limits occurs in the neighborhood of the critical densities (see equation 3.31), which are $N_c \approx 6 \times 10^2 \, \mathrm{cm}^{-3}$ for $^2D_{5/2}$ and $N_c \approx 3 \times 10^3 \, \mathrm{cm}^{-3}$ for $^2D_{3/2}$. The full solution of the equilibrium equations, which also takes into account all transitions, including excitation to the ^2P levels with subsequent cascading downward, gives the detailed variation of intensity ratio with the electron density that is plotted in Figure 5.3. Note from the collisional transition rates that the main dependence of this ratio is on $N_e/T^{1/2}$ and in addition there is a very slight temperature dependence (as a consequence of the cascading from ^2P) that cannot be seen on this graph.

An exactly similar treatment holds for [S II]; the calculated ratio $j_{\lambda6716}/j_{\lambda6731}$ is also shown in Figure 5.3. In [S II] the critical densities are higher than in [O II], so that the entire curve is shifted to somewhat higher densities. It is clear that the [O II] intensity ratio measures the electron density best in the neighborhood of $N_e \approx 10^3 \, \mathrm{cm}^{-3}$, while the [S II] ratio measures the density best around $N_e \approx 2 \times 10^3 \, \mathrm{cm}^{-3}$. Other ions with the same type of structure, which may also be used for measuring electron densities, are [Cl III], [Ar IV], and [K V].

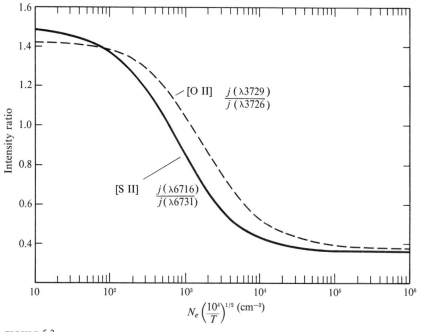

FIGURE 5.3
Calculated variation of [O II] (*solid line*) and [S II] (*dashed line*) intensity ratios as function of $N_e(10^4/T)^{1/2}$. Note that the electron density can be read directly from the horizontal scale if $T = 10,000°$K.

TABLE 5.6
Electron Densities in H II *Regions*

Object	$\dfrac{I(\lambda 3729)}{I(\lambda 3726)}$	$N_e \, (\mathrm{cm^{-3}})$
NGC 1976 A	0.50	4.5×10^3
NGC 1976 M	1.26	1.8×10^2
M 8 Hourglass	0.65	2.0×10^3
M 8 outer	1.26	2.0×10^2
NGC 281	1.37	9×10
NGC 7000	1.38	8×10

From the observational point of view, it is unfortunate that the [O II] $\lambda\lambda 3726$, 3729 are so close in wavelength, requiring a spectrograph, spectrometer, or interferometer with good wavelength resolution to separate the lines. However, a fair amount of data is available, both on H II regions and planetary nebulae.

Some results in H II regions, all derived from [O II] intensity ratios, are listed in Table 5.6. It must be remembered that only the brightest H II regions can be observed spectrophotometrically, so there is a selection in the data of Table 5.6 towards relatively high electron densities. Nevertheless, it can be seen that typical densities in several H II regions are of order $N_e \approx 10^2 \, \mathrm{cm^{-3}}$. (NGC 1976 M is a position in the outer part of the Orion Nebula.) Several H II regions have dense condensations in them, though—for instance, the central part of the Orion Nebula, near the Trapezium (NGC 1976 A), with $\lambda 3729/\lambda 3726 = 0.50$, corresponding to $N_e \approx 4.5 \times 10^3 \, \mathrm{cm^{-3}}$. In fact, observations of the [O II] ratio at many points in NGC 1976, of which only A and M are listed in Table 5.6, show that the mean electron density is highest near the center of the nebula, and decreases relatively smoothly outward in all directions. The three-dimensional structure of the nebula is presumably roughly spherically symmetrical, and the intensity ratio observed at the center results from emission all along the line of sight, so the actual central density must be higher than $4.5 \times 10^3 \, \mathrm{cm^{-3}}$. A model can be constructed that approximately reproduces all the measured [O II] ratios in NGC 1976; this model has $N_e \approx 1.7 \times 10^4 \, \mathrm{cm^{-3}}$ at the center and decreases to $N_e \approx 10^2 \, \mathrm{cm^{-3}}$ in the outer parts. Furthermore, measurements of the [S II] ratio at many points in the inner bright core of NGC 1976 (about 8′ diameter) show good agreement between the electron densities determined from the [S II] lines and the [O II] lines.

Similarly, in M 8 the [O II] measurements show that the density falls off outward from the Hourglass, a small dense condensation in which $N_e \approx 2 \times 10^3 \, \mathrm{cm^{-3}}$. Comparable published [S II] measurements do not exist for M 8, or indeed for any H II region except NGC 1976, but this situation should change rapidly in the next few years.

Some information on electron densities in planetary nebulae derived from [O II] and [S II] is listed in Table 5.7. It should be realized that in most planetaries the degree of ionization is high, and most of the [O II] and [S II] lines that arise in fairly low stages of ionization are emitted either in the outermost parts of the nebula, or the densest parts where recombination depresses the ionization the most. Thus the densities derived from these ions may not be representative of the entire nebula. The higher stages of ionization [Ar IV], [K V], and so on are better from this point of view, but their lines are weaker and more difficult to measure.

It should be noted that the collision strengths for [S II], which are less accurately calculated than those for [O II], have been adjusted to give overall agreement for all measured planetaries (essentially those of Table 5.7) between densities determined by [S II] and by [O II]. The best check is provided by the Orion Nebula measurements of [S II] and [O II] mentioned previously.

The electron densities derived from these line ratios may now be used in equations (5.4) and (5.5) to correct the observations of the tempera-

TABLE 5.7
Electron Densities in Planetary Nebulae

Nebula	[O II]		[S II]	
	$\dfrac{\lambda 3729}{\lambda 3726}$	$N_e/T^{1/2}$	$\dfrac{\lambda 6716}{\lambda 6731}$	$N_e/T^{1/2}$
NGC 40	0.78	14	—	—
NGC 650/1	1.23	2.3	—	—
NGC 2392	0.47	65	—	—
NGC 2440	0.64	27	—	—
NGC 3242	0.72	17	—	—
NGC 3587	1.34	1.2	—	—
NGC 6210	0.50	52	—	—
NGC 6543	0.50	52	0.60	65
NGC 6572	0.46	73	0.67	48
NGC 6720	1.01	5.8	—	—
NGC 6803	0.37	36	—	—
NGC 6853	1.16	3.3	—	—
NGC 7009	0.50	52	0.59	70
NGC 7027	0.48	60	0.60	65
NGC 7293	1.32	1.4	—	—
NGC 7662	0.56	45	0.62	60
IC 418	0.37	4×10^2	0.54	1×10^2
IC 2149	0.69	19	0.56	80
IC 4593	0.66	22	—	—
IC 4997	0.34	large	0.62	60

ture-sensitive lines of [O III] and [N II] for the slight collisional de-excitation effect, and actually these corrections have been taken into account in the results listed in Tables 5.1 and 5.2. Though the electron density derived from [O II] line measurements may not exactly apply in the [O III] emitting region, the density effect is small enough so that an approximate correction should be satisfactory.

In the densest planetaries known, collisional de-excitation of [O III] $^1D_2$ is strong enough so that the $(\lambda 4959 + \lambda 5007)/\lambda 4363$ ratio is significantly affected. The best example is IC 4997, with $\lambda 3729/\lambda 3726 = 0.35$, corresponding to N_e poorly determined in the high-density limit but certainly greater than 10^5 cm^{-3}. The measured $(\lambda 4959 + \lambda 5007)/\lambda 4363 \approx 22$, which would correspond to $T \approx 4 \times 10^4 \,^\circ\text{K}$ if there were no collisional de-excitation. This temperature is far too large to be understood from the known heating and cooling mechanisms, and the ratio is undoubtedly strongly affected by collisional de-excitation. If it is assumed that $T \approx 12,000 \,^\circ\text{K}$, the [O III] ratio gives $N_e \approx 10^6 \text{ cm}^{-3}$, while higher assumed temperatures correspond to somewhat lower electron densities and vice versa.

Many other ions have temperature-sensitive and electron-density-sensitive lines that are in the presently unobservable ultraviolet or infrared regions. As larger observatories above the earth's atmosphere become available, it will undoubtedly be very profitable to measure these lines in H II regions and in planetary nebulae.

5.6 Electron Temperatures and Densities from Radio Recombination Lines

Information can be obtained on the temperature and density in gaseous nebulae from measurements of the radio recombination lines. Practically all the observational results refer to H II regions, which have considerably larger fluxes than planetary nebulae and hence can be much more readily observed with radio telescopes. The populations of the high levels of H depend on T and N_e, as was explained in Section 4.4, and the relative strengths of the lines emitted by these levels with respect to the continuum and with respect to one another therefore depend on N_e, T, and the optical depth, which is conventionally expressed in terms of the emission measure E defined in equation (4.32). Comparison of measured and calculated relative strengths thus can be used to determine mean values of N_e, T, and E.

To calculate the expected strengths, we must solve the equation of radiative transfer, since the maser effect is often important, as was shown in Section 4.4. Furthermore, the continuum radiation is not weak in comparison with the line radiation and therefore must be included in the equation of transfer. The observations are generally reported in terms of brightness temperature, and we will use T_C for the measured brightness temperature in the continuum near the line and $T_L + T_C$ for the measured brightness temperature at the peak of the line (see Figure 5.4), so that T_L is the excess brightness temperature due to the line.

We shall consider an idealized homogeneous isothermal nebula; the optical depth in the continuum, which we shall write τ_C, is given by equation (4.32). The optical depth in the center of the line is

$$\tau_{cL} = \tau_L + \tau_C, \tag{5.10}$$

where τ_L is the contribution from the line alone,

$$d\tau_L = \kappa_L \, ds,$$

and

$$\kappa_L = N_n k_{OL}. \tag{5.11}$$

Here we consider an n, Δn line between an upper level $m = n + \Delta n$ and a lower level n; and the central line-absorption cross section, corrected for stimulated emission as in equation (4.43), is

$$k_{OL} = \frac{\omega_m}{\omega_n} \frac{\lambda^2}{8\pi^{3/2} \Delta\nu_D} A_{m,n} \left(1 - \frac{b_m}{b_n} e^{-h\nu/kT}\right)$$

$$= \frac{\omega_m}{\omega_n} \frac{\lambda^2 \ln 2}{4\pi^{3/2} \Delta\nu_L} A_{m,n} \left(1 - \frac{b_m}{b_n} e^{-h\nu/kT}\right). \tag{5.12}$$

In this equation a Doppler profile has been assumed, with $\Delta\nu_D$ the half width at e^{-1} of maximum intensity and $\Delta\nu_L$ the full width at half-maximum

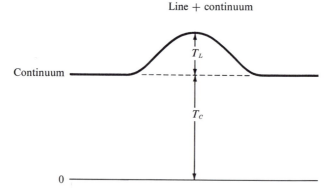

Line + continuum

FIGURE 5.4
A radio-frequency line superimposed on the radio-frequency continuum, showing the brightness temperatures at the center of the line and in the nearby continuum: T_L and T_C, respectively.

intensity, the conventional quantity used in radio astronomy. Combining (5.12) with

$$N_n = b_n n^2 \left(\frac{h^2}{2\pi m k T}\right)^{3/2} e^{X_n/kT} N_p N_e,$$
(5.13)

and using exp $(X_n/kT) \approx 1$ to a good approximation for all observed radio-frequency recombination lines, expressing $A_{m,n}$ in terms of the corresponding f-value f_{nm}, and expanding the stimulated-emission correction as in equation (4.43) gives, for the special case of local thermodynamic equilibrium $(b_m = b_n = 1)$, which we shall denote by an asterisk throughout this section,

$$\tau_L^* = 1.53 \times 10^{-9} \frac{n^2 f_{nm}}{\Delta \nu_L T^{2.5}} E_p$$

$$= 1.01 \times 10^7 \frac{\Delta n f_{nm}}{n \, \Delta \nu_L T^{2.5}} E_p.$$
(5.14)

The proton-emission measure

$$E_p = \int N_p N_e \, ds$$
(5.15)

is expressed in cm^{-6} pc in both forms of equation (5.14), and

$$\nu = \frac{\nu_0}{m^2} - \frac{\nu_0}{n^2} = \frac{2\nu_0 \Delta n}{n^3}.$$

In the true nebular case,

$$\tau_L = \tau_L^* b_n \frac{\left(1 - \frac{b_m}{b_n} e^{-h\nu/kT}\right)}{(1 - e^{-h\nu/kT})}$$

$$= \tau_L^* b_m \left(1 - \frac{kT}{h\nu} \frac{d \ln b_n}{dn} \Delta n\right),$$
(5.16)

while the continuum optical depth is the same as in thermodynamic equilibrium, because the free electrons have a Maxwellian distribution.

Now we will use these expressions and the formal solution of the equation of transfer to calculate the ratio of brightness temperatures $r = T_L/T_C$ in the special case of thermodynamic equilibrium,

$$r^* = \frac{T_L + T_C}{T_C} - 1 = \frac{T(1 - e^{-\tau_{CL}})}{T(1 - e^{-\tau_C})} - 1$$

$$= \frac{1 - e^{-(\tau_L^* + \tau_C)}}{1 - e^{-\tau_C}} - 1. \tag{5.17}$$

If $\tau_L^* \ll 1$ (this is a good approximation in all lines observed to date), and in addition, $\tau_C \ll 1$ (this is generally but not always a good approximation),

$$r^* = \frac{\tau_L^*}{\tau_C}.$$

Under the assumption of local thermodynamic equilibrium, the observed ratio of brightness temperatures in line and continuum thus gives (in the limit of small optical depth) the ratio of optical depths, which, in turn, from equations (4.32) and (5.14), measures T. Note that the continuum emission measure E involves all positive ions, but the proton emission measure E_p involves only H^+ ions, so their ratio depends weakly on the helium abundance, which, however, is reasonably well known. This scheme was used in the early days of radio recombination-line observations to determine the temperatures in H II regions, leading to values in the range 3000°–8000°, but it is not correct because in a nebula, the deviations from thermodynamic equilibrium are significant, as is shown by the fact that measurements of different lines in the same nebula, when reduced in this way, give different temperatures.

To calculate the brightness-temperature ratio $r = T_L/T_C$ in the true nebular case, we note that the brightness temperature in the continuum is still given by

$$T_C = T(1 - e^{-\tau_C}).$$

However, both the line-emission and line-absorption coefficients differ from their thermodynamic equilibrium values. The line-emission coefficient depends on the population in the upper level, so

$$j_L = j_L^* b_m,$$

while the line-absorption coefficient, as shown in equation (5.16), is

$$\kappa_L = \kappa_L^* b_m \beta,$$

where

$$\beta = 1 - \frac{kT}{h\nu} \frac{d \ln b_n}{dn} \Delta n. \tag{5.18}$$

The equation of transfer, in intensity units, is

$$\frac{dI_\nu}{d\tau_{LC}} = -I_\nu + \frac{j_L + j_C}{\kappa_L + \kappa_C} = -I_\nu + S_\nu, \qquad (5.19)$$

where

$$S_\nu = \frac{j_L^* b_m + j_C}{\kappa_L^* b_m \beta + \kappa_C}$$

$$= \frac{\kappa_L^* b_m + \kappa_C}{\kappa_L^* b_m \beta + \kappa_C} B_\nu(T) \qquad (5.20)$$

from Kirchoff's Law, so that the brightness temperature at the center of the line is

$$T_L + T_C = \left[\frac{\kappa_L^* b_m + \kappa_C}{\kappa_L^* b_m \beta + \kappa_C}\right] T[1 - e^{-(b_m \beta \tau_L^* + \tau_C)}]. \qquad (5.21)$$

Hence finally,

$$r = \frac{T_L}{T_C} = \left[\frac{\kappa_L^* b_m + \kappa_C}{\kappa_L^* b_m \beta + \kappa_C}\right]\left[\frac{1 - e^{-(b_m \beta \tau_L^* + \tau_C)}}{1 - e^{-\tau_C}}\right] - 1, \qquad (5.22)$$

which depends only on one optical depth, say, τ_C, the ratio of optical depths, $\tau_L^*/\tau_C = \kappa_L^*/\kappa_C$ given by equations (4.42) and (5.14), and the b_n factors, which, in turn, depend on N_e and T.

Thus, when the deviations from thermodynamic equilibrium are taken into account, r depends not only on T, but also on N_e and τ_C (or equivalently, E). Therefore, observations of several different lines in the same nebula are necessary to determine T, N_e, and E from measurements of radio-frequency recombination lines. The procedure is to make the best possible match between all measured lines in a given nebula, and the theoretical calculations for a given T, N_e, and E, using the $b_n(T, N_e)$ calculations described in Chapter 4. There are observational problems, connected with the fact that the measurements are made at different frequencies and with different radio telescopes, so the antenna beam patterns are not identical for all lines; but this does not seem to be a major source of error. Data on the best observed nebulae are collected in Table 5.8. Note that the last four nebulae in this table, W 3 through W 51, are all large distant H II regions, observed as bright sources in the radio region, but so strongly affected by interstellar extinction that they are completely unobservable in the optical region.

Finally, it should be noted that the overall fit between observed and calculated recombination line strengths is good for the lower quantum levels (say, those with $n \lesssim 100$ in all the nebulae observed), but for larger n, there

TABLE 5.8
Temperatures and Densities of Nebulae, from Radio Recombination-line Measurements

Nebula	$T(°K)$	$N_e \, (cm^{-3})$	$E \, (pc \, cm^{-6})$
NGC 1976	10,000	1.7×10^4	1.9×10^7
M 8	7700	4.9×10^3	1.0×10^6
M 17	7500	1.6×10^4	8.2×10^6
W 3	9500	7.3×10^4	4.5×10^7
W 43	7200	2.1×10^4	9.0×10^6
W 49	12,000	2.6×10^4	6.2×10^7
W 51	10,000	4.4×10^4	5.5×10^7

are discrepancies, even in the best possible fits, for several of the nebulae. These discrepancies are larger than the uncertainties of the calculations, and they indicate that the simplified uniform N_e, T models are not completely correct. More sophisticated models in which large-scale density variation in the nebula is taken into account give better overall agreement with the observed relative line strengths. The main new feature of these models is that the strong continuum radiation emitted in their dense central regions produces strong maser effects in the outer lower-density regions. The effects of small-scale, large-amplitude density fluctuations remain to be taken into account in the models. The study of radio recombination lines is very active at the present time, and undoubtedly new results will supersede some of the material in this section rather quickly.

5.7 Ionizing Radiation from Stars

Observations of gaseous nebulae may be used to find the numbers of ionizing photons emitted by a star and thus to determine a long base-line color index of a star, between the Lyman ultraviolet region and an ordinary optical region, from which the effective temperature of the star can be derived. The idea of the method is quite straightforward. If the nebula around the star is optically thick in the Lyman continuum, it will absorb all the ionizing photons emitted by the star. Thus the total number of ionizations in the nebula per unit time is just equal to the total number of ionizing photons emitted per unit time, and since the nebula is in equilibrium, these ionizations are just balanced by the total number of recaptures per unit time, so

$$\int_{\nu_0}^{\infty} \frac{L_\nu}{h\nu} \, d\nu = Q(H) = \int_0^{r_1} N_p N_e \alpha_B(H^0, T) \, dV,$$

where L_ν is the luminosity of the star per unit frequency interval. Note that by using the recombination coefficient α_B, we have included the ionization processes due to diffuse ionizing photons emitted in recaptures within the nebula—see equation (2.19). The luminosity of the entire nebula in a particular emission line, say, Hβ, also depends on recombinations throughout its volume:

$$L(\text{H}\beta) = \int_0^{r_1} 4\pi j_{\text{H}\beta}\, dV$$

$$= h\nu_{\text{H}\beta} \int_0^{r_1} N_p N_e \alpha_{\text{H}\beta}^{\text{eff}}(\text{H}^0, T)\, dV.$$

Thus dividing

$$\frac{\dfrac{L(\text{H}\beta)}{h\nu_{\text{H}\beta}}}{\displaystyle\int_{\nu_0}^{\infty} \frac{L_\nu}{h\nu}\, d\nu} = \frac{\displaystyle\int_0^{r_1} N_p N_e \alpha_{\text{H}\beta}^{\text{eff}}(\text{H}^0, T)\, dV}{\displaystyle\int_0^{r_1} N_p N_e \alpha_B(\text{H}^0, T)\, dV}$$

$$\approx \frac{\alpha_{\text{H}\beta}^{\text{eff}}(\text{H}^0, T)}{\alpha_B(\text{H}^0, T)} \tag{5.23}$$

gives us the result that the number of photons emitted by the nebula in a specific recombination line such as Hβ is directly proportional to the number of photons emitted by the star with $\nu \geq \nu_0$. Note that the proportionality between the number of ionizing photons absorbed and the number of line photons emitted does not depend on any assumption about constant density, and that replacing the ratio of integrals by the ratio of recombination coefficients is a good approximation because $\alpha_{\text{H}\beta}^{\text{eff}}/\alpha_B$ depends only weakly on T. Note further that any other emission line could have been used instead of Hβ, or alternatively, the radio-frequency continuum emission at any frequency at which the nebula is optically thin could have been used, except that then the ratio of nebular photons emitted to ionizing photons would involve the ratio of the number of protons to the total number of positive ions, which depends weakly on He abundance. The number of ionizing photons may be compared with the luminosity of the star at a particular frequency ν_f in the observable region,

$$\frac{L_{\nu_f}}{\displaystyle\int_{\nu_0}^{\infty} \frac{L_\nu}{h\nu}\, d\nu} = \frac{L_{\nu_f}}{\dfrac{L(\text{H}\beta)}{h\nu_{\text{H}\beta}}} \cdot \frac{\dfrac{L(\text{H}\beta)}{h\nu_{\text{H}\beta}}}{\displaystyle\int_{\nu_0}^{\infty} \frac{L_\nu}{h\nu}\, d\nu}$$

$$= h\nu_{\text{H}\beta} \frac{\alpha_{\text{H}\beta}^{\text{eff}}(\text{H}^0, T)}{\alpha_B(\text{H}^0, T)} \cdot \frac{\pi F_{\nu_f}}{\pi F_{\text{H}\beta}}, \tag{5.24}$$

where the ratio of luminosities has been expressed in terms of the ratio of the observed fluxes at the earth from the star at ν_f and from the nebula at Hβ. This ratio is independent of the distance, and is in addition independent of the interstellar extinction if the nebula and the star are observed at the same effective wavelength by choosing $\nu_f = \nu_{\text{H}\beta}$.

It is often more convenient to make the stellar measurements with a fairly wide filter of the type ordinarily used for photometry (for instance, the V filter of the UBV system), and we can then write a similar equation in terms of

$$L_V = \int_0^\infty s_\nu(V) L_\nu \, d\nu$$

and

$$\pi F_V = \int_0^\infty s_\nu(V) \pi F_\nu \, d\nu,$$

where $s_\nu(V)$ is the sensitivity function of the telescope-filter-photocell combination, known from independent measurements. For measurements of stars in bright nebulae, it is advantageous to use a narrower-band filter that isolates a region in the continuum between the brightest nebular emission lines, to minimize the correction for the "sky" background. In principle, any observable frequency ν_f can be used, and likewise any observable recombination line, for instance Hα, might be measured instead of Hβ. The method of using the nebular observations to measure the stellar ultraviolet radiation was first proposed by Zanstra, who assumed that the flux from a star could be approximately represented by the Planck function $B_\nu(T_*)$, so that

$$\frac{L_{\nu_f}}{\int_{\nu_0}^\infty \dfrac{L_\nu}{h\nu} \, d\nu} = \frac{B_{\nu_f}(T_*)}{\int_{\nu_0}^\infty \dfrac{B_\nu(T_*)}{h\nu} \, d\nu}$$

and the measurements thus determine T_*, the so-called Zanstra temperature of a star that ionizes a nebula. However, modern theoretical work on stellar atmospheres shows that there are important deviations between the emergent fluxes from stars and Planck functions, particularly in the regions where there are large changes in opacity with frequency, such as at the Lyman limit itself and at the various limits, due to other ions at shorter wavelengths, so that it is not a very good approximation to set $F_\nu = B_\nu(T_*)$. As illustrations, Figures 5.5 and 5.6 show calculated models for stars with $T_* = 40{,}000°$K, $\log g = 4$, approximately an O6 main-sequence star, and $T_* = 100{,}000°$K, $\log g = 6$, a fairly typical planetary-nebula star. Thus the ratios

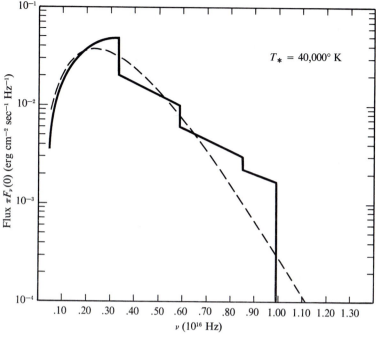

FIGURE 5.5
Calculated flux from a model O6 star with $T_* = 40,000°$K, $\log g = 4$ (*solid line*), compared with black-body flux for same temperature (*dashed line*).

$$\frac{L_{\nu_f}}{\int_0^\infty \dfrac{L_\nu}{h\nu}\,d\nu} = y(T_*) = \frac{\pi F_\nu(T_*, g)}{\int_{\nu_0}^\infty \dfrac{\pi F_\nu(T_*, g)}{h\nu}\,d\nu} \qquad (5.25)$$

should be determined from the best available sequences of model stellar atmospheres, and it can be seen that there is a one-parameter relationship $y = y(T_*)$ for a fixed value of g or along a fixed line in the T_*, $\log g$ plane.

We shall first use these relationships to examine the effective temperatures of Population I O stars in H II regions, and then generalize these equations and use them to describe the higher-temperature planetary-nebula central stars.

Many H II regions are observed, but a fairly large fraction of them contain several O stars that contribute to the ionization and thus complicate the determination of the effective temperature of individual stars. The best cases for measurement are clearly nebulae with only a single involved hot star. Furthermore, the basic assumption of the method is that the nebula completely absorbs the stellar ionizing radiation and is a true Strömgren sphere

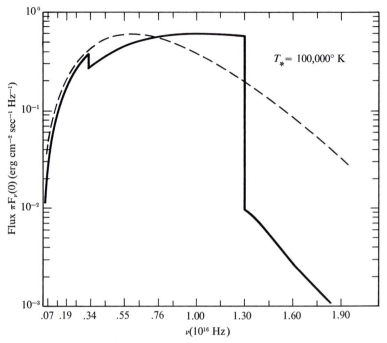

FIGURE 5.6

Calculated flux from a model planetary-nebula central star with $T_* = 100,000°K$, $\log g = 6$ (*solid line*), compared with black-body flux for same temperature (*dashed line*).

(radiation-bounded rather than density-bounded). It is difficult to be certain that this assumption is fulfilled in any particular nebula, though well-defined ionization fronts at the outer edge of a nebula suggest that it is and thus indicate that it is a good candidate to be measured.

There are not many available measurements of the total fluxes of nebulae in Hα or Hβ, but some photoelectric measurements do exist, and a somewhat larger number of photographic measurements have been made with narrow filter-plate combinations, particularly near Hα. These measurements have the defect that they include the [N II] λλ6548, 6583 lines within the filter band pass, but correction can be made for the contribution of these lines, at least statistically. There are considerably more nebulae for which radio-frequency continuum measurements are available, although they have the defect that interstellar extinction enters the ratio of optical stellar flux to radio-frequency nebular flux in full force. From these measurements of about 25 nebulae, the best available model stellar atmospheres have been used to derive the temperature scale for the main-sequence stars shown in Table 5.9.

In many planetary nebulae the number of ionizing photons emitted by the star beyond the He$^+$ limit can also be measured from $\lambda 4686$, the strongest He II recombination line. This line is too weak to be observed in any H II region, indicating that the flux of He$^+$-ionizing photons is small in all main-sequence O stars, confirming the calculated models in this respect. Many of the planetary-nebula central stars, however, are considerably hotter and emit an appreciable number of photons with $h\nu \geqslant 4h\nu_0 = 54.4$ eV. Thus, from the He II observations and from equation (2.29), we have the relation analogous to that in equation (5.24),

$$\frac{L_{\nu_f}}{\displaystyle\int_{4\nu_0}^{\infty} \frac{L_\nu}{h\nu}\,d\nu} = h\nu_{\lambda 4686}\,\frac{\alpha_{\lambda 4686}^{\text{eff}}(\text{He}^+, T)\pi F_{\nu_f}}{\alpha_B(\text{He}^+, T)\pi F_{\lambda 4686}}. \tag{5.26}$$

Hence from the measured H I and He II line fluxes of the nebula, together with the measured stellar flux at some observable frequency, two independent determinations of T_* can be made by the Zanstra method. In some nebulae these two determinations agree, but in other nebulae they disagree badly. For instance, in NGC 7662, the H I measurement indicates $T_* = 56{,}000°$K, while the He II measurement indicates that $T_* = 100{,}000°$K, which, in fact, corresponds to about 500 times more He$^+$-ionizing photons than does the lower temperature. The discrepancy may be understood as resulting from the fact that the nebula is not optically thick to the H-ionizing radiation as equation (5.24) assumes. If the nebula is density-bounded rather than ionization-bounded, then we must replace equation (5.24) with

$$\frac{L_{\nu_f}}{\displaystyle\int_{\nu_0}^{\infty} \frac{L_\nu}{h\nu}\,d\nu} = \eta_{\text{H}} h\nu_{\text{H}\beta}\,\frac{\alpha_{\text{H}\beta}^{\text{eff}}(\text{H}^0, T)\pi F_{\nu_f}}{\alpha_B(\text{H}^0, T)\pi F_{\text{H}\beta}}, \tag{5.27}$$

TABLE 5.9
Temperatures of Hot Main-sequence Stars Derived from Zanstra Method

Spectral type MK	$(B - V)_0$	T_*
O5	-0.32	$48{,}000°$
O6	-0.32	$40{,}000°$
O7	-0.32	$35{,}000°$
O8	-0.31	$33{,}500°$
O9	-0.31	$32{,}000°$
O9.5	-0.30	$31{,}000°$
B0	-0.30	$30{,}000°$
B0.5	-0.28	$26{,}200°$

where η_H represents the fraction of the H-ionizing photons that are absorbed in the nebula. Likewise, it is possible to imagine that all the He^+-ionizing photons are not absorbed within the nebula, that is, that even the He^{++} zone is density-bounded rather than ionization-bounded. However, this does not seem to occur in the observed planetaries because nearly all observed planetaries have He I lines in their observed spectra, indicating the existence of an outer He^+ zone, which, as the discussion of Chapter 2 shows, is certainly optically thick to He^+-ionizing radiation. In a similar way, if [O I] lines are observed in a nebula, they indicate the presence of O^0, and therefore also of H^0, which has the same ionization potential as O^0, and thus indicate that the nebula is optically thick to H-ionizing radiation and that $\eta_H = 1$.

One further item of information can be obtained from measurements of the flux in a He I recombination line, such as $\lambda 4471$ or $\lambda 5876$, namely, the number of photons emitted that can ionize He^0. This condition is:

$$\frac{L_{\nu_f}}{\int_{\nu_2}^{\infty} \frac{L_\nu}{h\nu} d\nu} = \eta_{He} h\nu_{\lambda 5876} \frac{\alpha_{\lambda 5876}^{eff}(He^0, T)\pi F_{\nu_f}}{\alpha_B(He^0, T)\pi F_{\lambda 5876}}. \tag{5.28}$$

If the nebula is known to be optically thick to the He-ionizing radiation, either because the He^+ zone is observed to be smaller than the H^+ zone, or because the apparent abundance ratio $N_{He^+}/N_p < 0.1$ (presumably indicating that the He^+ zone is smaller than the H^+ zone, even though this was not directly observed), then $\eta_{He} = 1$.

It should be noted that although the integrals giving the numbers of photons that can ionize He^+, He^0, and H^0 in equations (5.26), (5.28), and (5.27), respectively, overlap, the equations are nevertheless essentially correct, because, as indicated in Chapter 2, nearly every recombination of a He^{++} ion leads to emission of a photon that can ionize He^0 or H^0, and nearly every recombination of a He^+ ion leads to emission of a photon that can ionize H^0.

The observational data on the fluxes in $H\beta$, $\lambda 4686$, and $\lambda 4471$ are fairly complete and fairly accurate for planetary nebulae. The measurements of the stellar continuum fluxes are less accurate, because the stars are faint and must be observed on the bright background of the nebula.

Those planetary nebulae considered to be optically thick to H^0-ionizing radiation from the presence of [O I] lines in their spectra have Zanstra temperatures derived both from H I and from He I or He II lines in good agreement, a confirmation of the theory. For the other planetaries, T_* is derived from the He II measurements and equation (5.26), and the H I measurements are then used in equation (5.27) to calculate η_H. Some observational results are given in Table 5.10. The left-hand side of the table is

TABLE 5.10
Temperatures of Central Stars of Planetary Nebulae Derived from Zanstra Method

Nebula	T_*	
	Black-body approximation	Model atmosphere
NGC 2392	68,000	84,000
NGC 3242	93,000	109,000
NGC 3587	105,000	123,000
NGC 6210	50,000	—
NGC 6543	66,000	82,000
NGC 6572	61,000	77,000
NGC 6826	69,000	85,000
NGC 6853	132,000	148,000
NGC 7009	81,000	98,000
NGC 7662	100,000	118,000
IC 351	91,000	108,000
IC 418	43,000	—

based on the approximation $F_\nu(T_*, g) = B_\nu(T_*)$ for the fluxes of the central stars, while the right-hand side is based on values of πF_ν obtained from a series of model atmospheres calculated for planetary-nebula stars. Note that the calculated T_* are not very sensitive to the assumption, but the model atmospheres in general indicate somewhat higher temperatures.

5.8   Abundances of the Elements in Nebulae

It is clear that abundances of the observed ions in nebulae can be derived from measurements of the relative strengths of their emission lines. All the individual nebular lines are optically thin, so that no complicated curve-of-growth effects of the kind that complicate stellar atmosphere abundance determinations occur. Many light elements are observable in the nebulae, including H, He, N, O, and Ne, although unfortunately C is not. However, the strengths of collisionally excited lines depend strongly on temperature, which complicates the determination of relative abundances. Furthermore, all stages of ionization of an element are generally not observable in the optical spectral region; for instance, though [O II] and [O III] have strong lines in diffuse nebulae, O IV and O V do not.

In general, as we have seen in Chapter 4, the observed intensity I_l of an emission line is given by the integral

$$I_l = \int j_l \, ds = \int N_i N_e \epsilon_l(T) \, ds \qquad (5.29)$$

taken along the line of sight through the nebula, where N_i and N_e are the density of the ion responsible for the emission and the electron density, respectively.

For the recombination lines, the emission coefficients have been discussed in Chapter 4, and we have, for instance,

$$I_{H\beta} = \frac{1}{4\pi} \int N_p N_e h\nu_{H\beta} \alpha_{H\beta}^{\text{eff}} (H^0, T) \, ds,$$

$$I_{\lambda 5876} = \frac{1}{4\pi} \int N_{\text{He}^+} N_e h\nu_{\lambda 5876} \alpha_{\lambda 5876}^{\text{eff}} (He^0, T) \, ds,$$

$$I_{\lambda 4686} = \frac{1}{4\pi} \int N_{\text{He}^{++}} N_e h\nu_{\lambda 4686} \alpha_{\lambda 4686}^{\text{eff}} (He^+, T) \, ds.$$

For all the recombination lines, $\epsilon_l(T) \propto T^{-m}$ can be fitted over a limited range of temperature, with $m \approx 1$. For instance, for Hβ, $m = 0.90$, while for He I $\lambda 5876$, $m = 1.13$. Thus the recombination-emission coefficients are not particularly temperature sensitive and the abundances derived from them do not depend strongly on the assumed T.

Less abundant ions, such as C II, O IV, and O V, have weak permitted emission lines as observed in planetary nebulae, and these lines have often been interpreted as resulting from recombination, and have been used to derive abundances of the parent ions. However, much of the excitation of these lines is actually due to resonance-fluorescence, and their emission coefficients therefore depend not only on temperature and density but on the local radiation field as well, so actually they cannot be used to derive abundances in any straightforward way.

It is also possible to measure relative abundances of He$^+$ in H II regions from relative strengths of the radio recombination lines of H I and He I. At the very high n of interest in the radio region, both H and He are nearly identical one-electron systems except for their masses, so that the relative strengths of their lines (separated by the isotope effect) are directly proportional to their relative abundances as long as the lines are optically thin.

For abundance determinations of elements other than H and He, only collisionally excited lines are available, and for these lines, in contrast to the recombination lines, the emission coefficient depends more sensitively on the temperature,

$$I = \frac{1}{4\pi} \int N_i N_e \, h\nu \, q_{1,2}(T) \, b \, ds$$

$$= \frac{1}{4\pi} \int N_i N_e \, h\nu \, \frac{8.63 \times 10^{-6}}{T^{1/2}} \frac{\Omega(1,2)}{\omega_1} e^{-\chi/kT} b \, ds,$$

in the low-density limit, where b is the fraction of excitations to level 2 that are followed by emission of a photon in the line observed.

The temperature must be determined from observational data of the kind discussed in the beginning sections of this chapter. From the measured relative strengths of the lines and the known emission coefficients, the abundances can be determined on the basis of a model of the structure of the nebula. The simplest model treats the nebula as homogeneous with constant T and N_e and thus might be called a one-layer model. From each observed relative line strength, the abundance of the ion that emits it can be determined. In some cases two successive stages of ionization of the same element are observed, such as O^+ and O^{++}, and their relative abundances can be used to construct an empirical ionization curve giving $N(A^{+m+1})/N(A^{+m})$ as a function of ionization potential. Thus finally, the relative abundance of every element with at least one observed line can be determined. Discrepancies (for instance, in N_e and T) determined from different line ratios indicate that this model is too simplified to give highly accurate results, though the abundances determined from it are generally thought to be correct to within a factor of order two or three.

A somewhat more sophisticated scheme takes into account the spatial variations of temperature and uses the observations themselves to get as much information as possible on these variations. The emission coefficient is expanded in a power series

$$\epsilon_l(T) = \epsilon_l(T_0) + (T - T_0)\left(\frac{d\epsilon_l}{dT}\right)_0 + \frac{1}{2}(T - T_0)^2 \left(\frac{d^2\epsilon_l}{dT^2}\right)_0, \quad (5.30)$$

correct to the second order. It is clear that for recombination lines with

$$\epsilon_l(T) = CT^{-m}$$

or for collisionally excited lines with

$$\epsilon_l(T) = \frac{De^{-\chi/kT}}{T^{1/2}},$$

the necessary derivatives can be worked out analytically. Then integrating along the line of sight,

$$\int N_i N_e \, \epsilon_l(T) \, ds$$

$$= \epsilon_l(T_0) \int N_i N_e \, ds + \frac{1}{2}\left(\frac{d^2\epsilon_l}{dT^2}\right)_0 \int N_i N_e (T - T_0)^2 \, ds, \quad (5.31)$$

where T_0 is chosen so that

$$T_0 = \frac{\int N_i N_e \, T \, ds}{\int N_i N_e \, ds}. \tag{5.32}$$

If all ions have the same space distribution $N_i(s)$, then from two line ratios, such as [O III] ($\lambda 4959 + \lambda 5007$)/$\lambda 4363$ and [N II] ($\lambda 6548 + \lambda 6583$)/$\lambda 5755$, both T_0 and

$$t^2 = \frac{\int N_i N_e (T - T_0)^2 \, ds}{T_0^2 \int N_i N_e \, ds}$$

can be determined instead of the one constant T_0 from one line ratio, as in the single-layer model. Then T_0 and t^2 can be used to determine the abundances of all the ions with measured lines. The difficulty with this method is that all ions do not have the same space distribution; for instance, O^{++} is more strongly concentrated to the source of ionizing radiation than N^+, so other more or less arbitrary assumptions must be made.

The most sophisticated method of all to determine the abundances from the observations is to calculate a complete model of the nebula in an attempt to reproduce all its observed properties; this approach will be discussed in the next section.

Turning now to the observational results, the He/H abundance ratio has been measured in many nebulae. Perhaps the most exhaustively measured nebula is the Orion Nebula, for which the most recent results are $N_{He^+}/N_p = 0.097$, 0.095, 0.077, and 0.009 in four slit positions. The last measurement is not in NGC 1976 itself, but in the nearby companion nebula NGC 1982, and it definitely shows that that slit position is in an H^+, He^0 zone where He is neutral. The exciting star of NGC 1982 is a B1 V star, so the fact that the nebula is a He^0 zone is understood from Figure 2.5. This observation shows that some correction of the abundance of He for the unobserved He^0 is probably necessary at all the observed slit positions in the Orion Nebula. Empirically, the correction can be based on the observed strength of [S II] $\lambda\lambda 6717, 6731$, because their emitting ion S^+ has an ionization potential of 23.4 eV, approximately the same as the ionization potential of He^0, 24.6 eV, so that, to a first approximation,

$$\frac{N_{He^0}}{N_{He^+}} = \frac{N_{S^+}}{N_{S^{++}}}.$$

A slightly better empirical determination can be made by interpolating between the relative abundance of S^+ (ionization potential 23.4 eV) and

TABLE 5.11

Comparison of Optical and Radio Helium Abundance Determinations

Nebula	He+/H+ optical	He+/H+ radio
NGC 1976	0.090	0.080
NGC 6618	0.097	0.086

O^+ (ionization potential 35.1 eV) to find the abundance of He^0. The final result for NGC 1976 is $N_{He}/N_H = 0.11$; two other H II regions observed optically, M 8 and M 17, have essentially this same relative He abundance.

Radio measurements of He+/H+ abundance ratios are available for many diffuse nebulae. These determinations are in fairly good agreement with the optical measurement for nebulae common to both sets of observations, as shown in Table 5.11. The average abundance ratio from seven H II regions observed in the radio-frequency region is $N_{He^+}/N_p = 0.08$. At present, however, there is no known way in which the correction for He^0 can be obtained from radio measurements alone, and the fact that at least two nebulae, NGC 2024 and NGC 1982, are observed to have $N_{He^+}/N_p \approx 0$ shows that this correction certainly exists. The radio measurements open the fascinating possibility of determining relative He/H abundances at very great distances in the Galaxy (for instance, in H II regions near the galactic center) that are unobservable optically because of interstellar extinction.

Both He II and He I recombination lines are observed in many planetary nebulae, showing the presence of both He^{++} and He^+, though some planetaries, like H II regions, have only He I lines. Nearly all planetaries have central stars that are so hot that they have no outer H^+, He^0 zones, though a few exceptions have been discussed in previous sections of this text. Thus no correction is necessary for unobserved He^0 in most planetary nebulae.

If only the most accurate photoelectric measurements of planetary nebulae are used, the derived helium abundances have only a small scatter, as shown in Table 5.12. The average for the ten best observed planetaries in this table

TABLE 5.12

Helium Abundances in Planetary Nebulae

Nebula	N_{He^+}/N_p	$N_{He^{++}}/N_p$	N_{He}/N_H
NGC 1535	0.09	0.01	0.10
NGC 6572	0.11	0.00	0.11
NGC 6720	0.07	0.04	0.11
NGC 6803	0.12	0.00	0.12
NGC 6884	0.11	0.01	0.12
NGC 7009	0.10	0.01	0.11
NGC 7027	0.08	0.04	0.12
NGC 7662	0.05	0.05	0.10
IC 418	0.07	0.00	0.10
IC 5217	0.09	0.01	0.10

is $N_{He}/N_H = 0.11$, while the complete range is from 0.10 to 0.12. Only in IC 418 was a correction for the He^0 necessary. Since the accuracy of the measurements, as judged from the relative intensities of He I $\lambda\lambda4471, 5876$, is about 0.01, the differences between the nebulae are marginal in any case and may not exist at all. Some special planetaries with well-determined helium abundances are K 648 in the globular cluster M 15 with $N_{He}/N_H = 0.13$ and 49 $+88°$ 1 near the galactic pole, approximately 20 kpc distant, with $N_{He}/N_H = 0.13$. Thus these extreme Population II objects have helium abundances that are indistinguishable from the more commonly observed bright planetary nebulae near the sun.

Among H II regions, the most complete abundance determinations of the heavy elements are available for NGC 1976. The results given in Table 5.13 show that it has fairly normal abundances of N, O, Ne, and S. Similar heavy-element abundance determinations are available for M 8 and M 17, with results nearly the same as for NGC 1976. Abundance determinations have been made for a larger number of planetary nebulae, and the average results for these planetaries, taken from a recent summary paper, are also shown in Table 5.13. Still more recent and accurate measurements show that though most elements have normal abundances in planetary nebulae, N is somewhat overabundant. The main difficulty with the abundance determinations for the heavy elements is that large and rather uncertain corrections are required for unseen ions, that is, ions without observable lines in the optical region. For this reason, the calculations of model planetary nebulae that are described in the next section probably represent the most nearly accurate method for finding abundances of the heavy elements.

TABLE 5.13

Abundances of Elements in Gaseous Nebulae

Element	Logarithm of relative abundance	
	Average planetary nebula	NGC 1976
H	12.00	12.00
He	11.23	11.04
N	8.1:	7.63
O	8.9	8.79
F	4.9	—
Ne	7.9	7.86
Na	6.6	—
S	7.9	7.50
Cl	6.9	—
Ar	7.0	—
K	5.7	—
Ca	6.4:	—

5.9   Calculations of the Structure of Model Nebulae

The basic idea of a calculation of a model H II region or a model planetary nebula is quite straightforward. It is to make reasonable assumptions about the physical parameters of the ionizing star, the density distribution, and the relative abundances of the elements in the nebula (its size, geometrical structure, and so on); to calculate, on the basis of these assumptions, the resulting complete physical structure—the ionization, temperature, and emission coefficients as functions of position; and thus to calculate the expected emergent radiation from the nebula at each point in each emission line. Comparing this predicted model with the observed properties of a nebula provides a check as to whether the initial assumptions are consistent with the observations; if they are not, then the assumptions must be varied until a match with the observational data is obtained. In principle, if all the emission lines were accurately measured at every point in the nebula, and if the central star's radiation were measured at each observable frequency, it might be possible to specify accurately all the properties of the star and of the nebula in this way. Of course, in practice the observations are not sufficiently complete and accurate, and do not have sufficiently high angular resolution to enable us to carry out this ambitious program, but nevertheless, quite important information is derived from the model-nebula calculation.

Let us write down in simplified form the equations used in calculating the structure of a model nebula. For computational reasons, practically all work to date has assumed spherical model nebulae, and we shall write the equations in these terms. The basic equations are described in Chapters 2 and 3, so we shall simply quote them here. The equation of transfer is

$$\frac{dI_\nu}{ds} = -\frac{d\tau_\nu}{ds} I_\nu + j_\nu, \tag{5.33}$$

where the increment in optical depth at any frequency is given by a sum

$$\frac{d\tau_\nu}{ds} = \sum N_j a_{\nu_j} \tag{5.34}$$

over all atoms and ions with ionization potentials $h\nu_j < h\nu$. In practice, because of their great abundance, only H^0, He^0, and He^+ need be taken into account except possibly at the very highest frequencies. Likewise, the emission coefficient j_ν is a sum of terms of which again only those due to recombinations of H^+, He^+, and He^{++} are important.

The ionization equation that applies between any two successive stages of ionization of any ion is

$$N(X^{+i}) \int_{\nu_i}^{\infty} \frac{4\pi J_\nu}{h\nu} a_\nu(X^{+i}) \, d\nu = N(X^{+i+1}) N_e \alpha_A(X^{+i}, T) \qquad (5.35)$$

as in equation (2.30), while the total number of ions in all stages of ionization is

$$\sum_{i=0}^{\text{max}} N(X^{+i}) = N(X).$$

The energy-equilibrium equation is

$$G = L_R + L_{FF} + L_C \qquad (5.36)$$

as in equation (3.33), where the gain term and each of the loss terms is a sum over the contributions of all ions, but again in practice only H, He, and He$^+$ need be taken into account in L_R and L_{FF}. Collisionally excited line radiation from the less abundant heavy elements dominates the cooling, however, and many terms must be included in L_C.

For any assumed input-radiation source at the origin, taken to be a star with either a black-body spectrum, or a spectrum calculated from a model stellar atmosphere, these equations can be integrated. If the on-the-spot approximation described in Chapter 2 is used, they can be integrated outwards. If instead, the detailed expressions for the emission coefficients are used and the diffuse radiation field is explicitly calculated, it is necessary to use an iterative procedure. The on-the-spot approximation can be used as a first approximation from which the ionization at each point in the nebula and the resulting emission coefficients can be calculated. Then the diffuse radiation field can be calculated working outward from the origin, and using the then more nearly accurate total radiation field, the ionization and T can be recalculated at each point. This process can be repeated as many times as needed until it converges to the desired accuracy.

As an example, we shall examine the available models of what is probably the most studied planetary nebula, NGC 7662, a bright, northern object with a highly symmetric double-ring structure. Three different sets of models have been calculated for this nebula by three different authors. Their most accurately assumed input parameters and the resulting calculated relative intensities (for the whole nebula) are listed in Table 5.14. It can be seen from the table that the agreement between the observed data and the calculated models is reasonably good, but not perfect.

All the authors found that, with assumed homogeneous models, if the observed strengths of the lines from the higher stages of ionization, such

TABLE 5.14
Models of NGC 7662

	Harrington 1969	Flower 1969	Kirkpatrick 1972	Observed (corrected for extinction)
Distance (pc)	1000	1530	830	
Star				
T_*	100,000°	100,000°	95,090°	
	Black-body model	Model $\log g = 5$	Model with extended atmosphere	
$L/L_\odot$	5.1×10^3	1.0×10^4	1.7×10^4	
Nebula				
Inner radius (pc)	0.030	0.030	0.020	
Outer radius (pc)	0.072	0.090	0.067	
N_H (maximum density outside of condensations, cm^{-3}	2.0×10^3	7.0×10^3	1.3×10^4	
Condensations	Spherical—with density 5 times mean density; occupy 0.03 of volume of outer part of nebula	Spherical—occupy 0.09 of volume of outer part of nebula, rest is vacuum	None—continuous range of density from 10 cm^{-3} to 1.4×10^4 cm^{-3}	
Relative abundances				
H	1.00	1.00	1.00	
He	0.16	0.16	0.19	
N	2.5×10^{-4}	5.5×10^{-5}	1.5×10^{-4}	
O	4.5×10^{-4}	2.5×10^{-4}	2.1×10^{-4}	
Ne	5.5×10^{-5}	6.3×10^{-5}	4.5×10^{-5}	
S	—	—	1.2×10^{-5}	
Ar	—	—	1.8×10^{-6}	
Calculated line strengths ($H\beta = 100$)				
He I 5876	17	—	12	7
He II 4686	61	36	69	43
[N II] 6548 + 6583	6.3	3.9	2.3	12
[O II] 3726 + 3729	7.1	7.2	13	13
[O III] 4363	18	15	19	16
[O III] 4959 + 5007	1810	1315	1240	1465
[Ne III] 3869 + 3967	107	117	125	110
[Ne V] 3346 + 3425	17	60	22	17

as [O III], [Ne III], [Ne V], were matched by the model, then the lines resulting from lower stages, such as [O II] and [N II], were calculated to be considerably weaker than observed. Therefore, in two cases, the final model assumed was homogeneous in the inner parts, but with some type of density condensations in the outer parts, to enhance the emission from the lowest stages of ionization. In the first model, the small dense condensations are assumed to be immersed in a lower-density medium that occupies most of the volume in the outer parts of the nebula, while in the second model, the small dense condensations are assumed to lie in a vacuum. In fact, the direct photographs show that there are condensations in the outer parts of this nebula, so both models do provide a rough approximation of its real structure. The third model has no condensations, but instead assumes a spherical density distribution that is approximately the sum of two Gaussian functions, and thus simulates the double-shell appearance of NGC 7662. Note from Table 5.14, however, that the calculated strengths of the [O II] and [N II] lines remain somewhat weaker than the observed strengths, even though condensations or density variations have been assumed in all three models. Models of other planetaries also have this same property, and generally also predict the [O I] lines to be weaker than their observed strengths. Evidently, the models do not completely represent the actual structure at the outer edge of the nebula. This is perhaps due to the fact that all the models are calculated equilibrium ionization-bounded structures, while in the outermost parts this cannot be a correct assumption. This point will be discussed further in the following chapter on nebular dynamics.

None of the models listed in Table 5.14 was calculated taking into account the charge-exchange processes described in Section 2.8. However, more recently calculated models, which do include these processes but are otherwise identical with the Flower model, give much better agreement between the calculated and observed [O I] line strengths. This result confirms that charge-exchange is important in the outer parts of some nebulae, and should be included in future model calculations.

Models of the type discussed here have also been applied to H II regions. They are apparently the most realistic available representations of these highly chaotic objects, but there are few observations of the optical-line radiation with which they may be compared. The fundamental problem, however, is that the assumption of spherical symmetry, and in particular of homogeneity, or else of a density distribution that depends only on distance from the center, possibly including some simplified form of density fluctuations, is far too simple to use in describing real nebulae. Though some planetaries have a fairly symmetric form and smooth structure, no H II regions do. The photographs in this book show the very complicated structure of actual nebulae, and the example of the planetary nebula NGC 6853 in Figure 5.7 demonstrates how strongly the local structure can affect

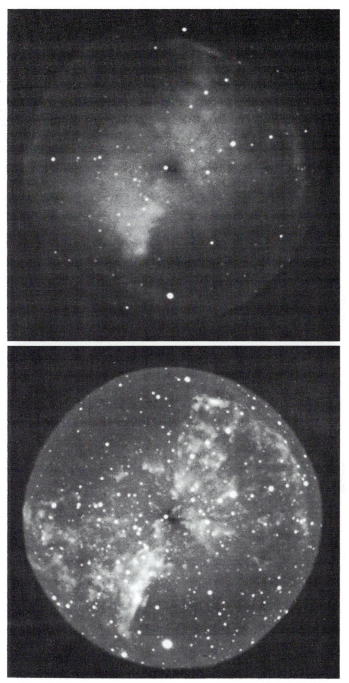

FIGURE 5.7
Monochromatic photographs of the planetary nebula NGC 6853: in the
light of Hβ λ4861 (*top*), and in the light of [O I] λ6300 (*bottom*). Note
that, in contrast to the Hβ emission, the [O I] emission is strongly
concentrated to many bright spots, which must be high-density neutral
condensations (surrounded by partly ionized edges) in which both O^0
and free electrons are present. (*Steward Observatory photographs.*)

line-emission coefficients. It is clear that models that represent this structure in some realistic way will be necessary before the nebulae can be considered to be adequately represented.

References

A very good overall reference on the comparison of theory and observation of the optical radiation of nebulae, written in the context of planetary nebulae but applicable in many ways to H II regions also is:

Seaton, M. J. 1960. *Rep. Progress in Phys.* 1960 **23**, 313.

The method of measuring electron temperatures from optical emission-line intensity ratios seems to have been first suggested by

Menzel, D. H., Aller, L. H., and Hebb, M. H. 1941. *Ap. J.* **93**, 230.

This method has subsequently been used by many authors.

The form of the [O III] ratio plotted in Figure 5.1, which includes the effects of the resonances on the collision strengths Ω, is from

Eissner, W., Martins, P. de A. P., Nussbaumer, H., Saraph, H. E., and Seaton, M. J. 1969. *M. N. R. A. S.* **146**, 63.

The observational data used in Table 5.1 are from

Peimbert, M., and Costero, R. 1969. *Bol. Obs. Tonantzintla Tacubaya* **5**, 3.

The data in Table 5.2 are from

O'Dell, C. R. 1963. *Ap. J.* **138**, 1018.

Peimbert, M., and Torres-Peimbert, S. 1971. *Bol. Obs. Tonantzintla Tacubaya* **36**, 21.

The observational data for the optical continuum temperature determinations of Table 5.4 are from

Peimbert, M. 1971. *Bol. Obs. Tonantzintla Tacubaya* **36**, 29.

The data for NGC 7027 described in the text of Section 5.3 are from

Miller, J. S., and Mathews, W. G. 1972. *Ap. J.* **172**, 593.

The radio-continuum method of measuring the temperature of a nebula by observing it in the optically thick region has been discussed and used by many authors. For example:

Wade, C. M. 1958. *Austr. J. Phys.* **11**, 388.

Wade, C. M. 1959. *Austr. J. Phys.* **12**, 418.

Menon, T. K. 1961. *Pub. Nat. Radio Astron. Obs.* **1**, 1.

Mills, B. Y., Little, A. G., and Sheridan, K. V. 1956. *Austral. J. Phys.* **9**, 218.
Terzian, Y., Mezger, P. G., and Schraml, J. 1968. *Ap. Letters* **1**, 153.
Shaver, P. A. 1970. *Ap. Letters* **5**, 167.
(Table 5.5 is based on observational material from this last reference.)
Thompson, A. R. 1967. *Ap. Letters* **1**, 25.
Le Marne, A. E., and Shaver, P. A. 1969. *Proc. Austral. Soc. Astron.* **1**, 216.

The most complete information on radio-frequency measurements of planetary nebulae, including the most accurate temperatures derived by this method, is included in

Higgs, L. A. 1971. *Catalog of Radio Observations of Planetary Nebulae and Related Optical Data.* National Research Council of Canada Publication NRC 12129.
Higgs, L. A. 1973. *M. N. R. A. S.* **161**, 305.

The observational data on the small "hot spot" in NGC 7027 detected by 11-cm interferometer measurements are from

Miley, G. K., Webster, W. J., and Fullmer, J. W. 1970. *Ap. Letters* **6**, 17.
The 6-cm and 11-cm measurements of lower angular resolution that do not show it are included in

Webster, W. J., Wink, J. E., and Altenhoff, W. J. 1970. *Ap. Letters* **7**, 47.
Scott, P. F. 1973. *M. N. R. A. S.* **161**, 35P.

The idea of using the [O II] intensity ratio to measure electron densities in nebulae seems to have been first suggested by

Aller, L. H., Ufford, C. W., and Van Vleck, J. H. 1949. *Ap. J.* **109**, 42.
It was worked out quantitatively by

Seaton, M. J. 1954. *Ann. d'Astrophys.* **17**, 74.

A complete discussion, including theoretical calculations and also observational data on several of the H II regions used in Table 5.6, is found in

Seaton, M. J., and Osterbrock, D. E. 1957. *Ap. J.* **125**, 66.
The most recent and best calculation of the variation of the [O II] ratio with N_e, taking account of the effect of resonances, is included in

Eissner, W., Martins, P. de A. P., Nussbaumer, H., Saraph, H. E., and Seaton, M. J. 1969. *M. N. R. A. S.* **146**, 63.
(Figure 5.3 is based on this reference.)

Note that the calculated transition probabilities, which are very sensitive to the exact departures from *LS* coupling, give a calculated high-density limit of the ratio $I(\lambda 3729)/I(\lambda 3726) = 0.45$ that is higher than the ratio observed in several planetary nebulae. Therefore, the transition probabilities have been adjusted by Eissner *et al.* to bring the calculated high-density limit down to $I(\lambda 3729)/I(\lambda 3726) = 0.35$, in the densest observed planetary, IC 4997. (These adjusted transition probabilities are the ones listed in Table 3.9.)

The observational data in Tables 5.6 and 5.7 and also the data discussed in Section 5.5 are from

Osterbrock, D., and Flather, E. 1959. *Ap. J.* **129**, 26. (NGC 1976).

Osterbrock, D. E. 1960. *Ap. J.* **131**, 541. (Planetary nebulae).

Meaburn, J. 1969. *Astron. and Space Science* **3**, 600. (M 8).

Danks, A. C. 1970. *Astron. and Space Science* **9**, 175. [S II].

Danks, A. C., and Meaburn, J. 1971. *Astron. and Space Science* **11**, 398. (NGC 1976).

Comparison of radio-recombination-line measurements with calculated strengths and the determination of T, N_e, and E have been investigated by many authors, and the treatment in the present text is based on

Goldberg, L. 1966. *Ap. J.* **144**, 1225.

Mezger, P. G., and Hoglund, B. 1967. *Ap. J.* **147**, 490.

Dyson, J. E. 1967. *Ap. J.* **150**, L45.

Hjellming, R. M., Andrews, M. H., and Sejnowski, T. J. 1969. *Ap. J.* **157**, 573.

Dupree, A. K., and Goldberg, L. 1970. *Ann. Rev. Astr. and Astrophys.* **8**, 231.

Hjellming, R. M., and Davies, R. D. 1970. *Astron. and Astrophys.* **5**, 53.

Hjellming, R. M., and Gordon, M. A. 1971. *Ap. J.* **164**, 47.

Andrews, M. H., Hjellming, R. M., and Churchwell, E. 1971. *Ap. J.* **167**, 245.

Note, however, that the definition of β used in the present text differs slightly from that used in most of these papers. (Table 5.8 is based on the last two of these references.)

The importance of density variations in the models was described and illustrated by

Brocklehurst, M., and Seaton, M. J. 1971. *Ap. Letters* **9**, 139.

Brocklehurst, M., and Seaton, M. J. 1972. *M. N. R. A. S.* **157**, 179.

Very complete lists of f-values for radio-recombination lines are given by

Goldwire, H. C. 1968. *Ap. J. Supp.* **17**, 445.

Menzel, D. H. 1969. *Ap. J. Supp.* **18**, 221.

The method of measuring the ultraviolet radiation of stars from the recombination radiation of the nebulae they ionize was first suggested by

Zanstra, H. 1931. *Pub. Dominion Astrophys. Obs.* **4**, 209.

The treatment in this chapter is based primarily on the following modern papers, which include the best available optical and radio-frequency measurements:

Harman, R. J., and Seaton, M. J. 1966. *M. N. R. A. S.* **132**, 15.

(The black-body temperature determinations of Table 5.10 are taken from this reference.)

Capriotti, E. R. 1967. *Ap. J.* **147**, 979.

Gebel, W. L. 1968. *Ap. J.* **153**, 743.

Capriotti, E. R., and Kovach, W. S. 1968. *Ap. J.* **151**, 991.

(The model stellar-atmosphere determinations of Table 5.10 are taken from this reference.)

Morton, D. C. 1969. *Ap. J.* **158,** 629.
(The temperature determinations of Table 5.9 are taken from this reference.)

Recent surveys of abundance determinations in nebulae include
Aller, L. H., and Czyzak, S. J. 1968. *Planetary Nebulae* (IAU Symposium No. 34). Dordrecht: D. Reidel, p. 209
(the average data of Table 5.13 are taken from this reference) and
Osterbrock, D. E. 1970. *Q. J. R. A. S.* **11,** 199.
The radio determinations of helium abundances of Table 5.11 are from
Palmer, R., Zuckerman, B., Penfield, H., Lilley, A. E., and Mezger, P. G. 1969. *Ap. J.* **156,** 887.
Reifenstein, E. C., Wilson, T. L., Burke, B. F., Mezger, P. G., and Altenhoff, W. J. 1970. *Astron. and Astrophys.* **4,** 357.
Two very recent publications, in which much lower helium abundances are reported in H II regions near the galactic center, are:
Mezger, P. G., and Churchwell, E. B. 1973. *Nature* **242,** 319.
Huchtmeier, W. K., and Batchelor, R. A. 1973. *Nature* **243,** 155.

The optical determinations of the helium abundances in Tables 5.11 and 5.12 are from
Peimbert, M., and Costero, R. 1969. *Bol. Obs. Tonantzintla Tacubaya* **5,** 3. (H II regions).
Peimbert, M. and Torres-Peimbert, S. 1971. *Ap. J.* **168,** 413. (Planetary nebulae).
An earlier, very complete discussion of helium in planetary nebulae is included in
Harman, R. J., and Seaton, M. J. 1966. *M. N. R. A. S.* **132,** 15.

Among the many discussions of complete models of planetary nebulae, some of the best are:
Harrington, J. P. 1968. *Ap. J.* **152,** 943.
Harrington, J. P. 1969. *Ap. J.* **156,** 903.
Flower, D. R. 1969. *M. N. R. A. S.* **146,** 171.
Flower, D. R. 1969. *M. N. R. A. S.* **146,** 243.
Kirkpatrick, R. C. 1970. *Ap. J.* **162,** 33.
Kirkpatrick, R. C. 1972. *Ap. J.* **176,** 381.
(The data of Table 5.14 are taken from the second, fourth, and sixth of these references.)

Model H II regions are systematically discussed by
Hjellming, R. M. 1966. *Ap. J.* **143,** 420.
Rubin, R. H. 1968. *Ap. J.* **153,** 761.

6

Internal Dynamics of Gaseous Nebulae

6.1 Introduction

The first five chapters of this book have described gaseous nebulae entirely from a static point of view. However, this description cannot be complete, because nebulae certainly have internal motions, and the effects of these motions on their structures cannot be ignored. It can easily be seen that an ionized nebula cannot be in static equilibrium, for if it is density-bounded, it will expand into the surrounding vacuum, while if it is ionization-bounded, the hot ionized gas (with $T \approx 10,000\,°\mathrm{K}$) will initially have a higher pressure than the surrounding cooler neutral gas ($T \approx 100\,°\mathrm{K}$) and will therefore tend to expand until its density is low enough so that the pressures of the two gases are in equilibrium. In addition, when the hot star in a nebula first forms and the source of the ionizing radiation is thus "turned on," the ionized volume initially grows in size at a rate fixed by the rate of emission of ionizing photons, and an ionization front separating the ionized and neutral regions propagates into the neutral gas.

Observations agree in showing that the internal velocities of nebulae are not everywhere zero. Measured radial velocities show that planetary nebulae are expanding more or less radially; mean expansion velocities are of order 25 km sec^{-1}, and the velocity gradient is positive outward. Many H II regions are observed to have chaotic internal velocity distributions that can best be described as turbulent.

This chapter will therefore concentrate on the internal dynamics of nebulae. First it considers the hydrodynamic equations of motion that are applicable to nebulae, particularly in the spherically symmetric form in which these equations have actually been applied to date. This discussion leads to a study of ionization fronts and of shock fronts that are generated by ionization fronts and by the expansion of the nebulae. Then all the available theoretical results for planetary nebulae and H II regions are analyzed. Finally, a brief synopsis of the available observational material is given, and it will be seen that much more theoretical work is necessary before the observations can be fully understood, or even described in an optimum way.

6.2 Hydrodynamic Equations of Motion

The standard equation of motion for a compressible fluid, such as the gas in a nebula, may be written

$$\rho \frac{D\mathbf{v}}{Dt} \equiv \rho \left(\frac{\partial \mathbf{v}}{\partial t} + \mathbf{v} \cdot \nabla \mathbf{v} \right) = -\nabla p - \rho \, \nabla \phi, \qquad (6.1)$$

where D stands for the time derivative following an element of the gas, and ∂ stands for the partial time derivative at a fixed point in space. On the right-hand side of the equation, the forces included are the pressure gradient and the force resulting from the gravitational potential of the involved stars and of the nebula itself. However, the dimensions of any observed structure in a nebula are so large that the gravitational forces are negligibly small, and the second term can be omitted. Note, however, that in equation (6.1) electromagnetic forces have also been omitted; there may be nebulae in which magnetic fields exist that are sufficiently large so that this omission is incorrect, but as there is no strong evidence for the existence of these fields and since they would complicate the problem enormously, we shall make this simplification. The gas pressure is

$$p = \frac{\rho k T}{\mu m_{\mathrm{H}}}, \qquad (6.2)$$

and in most situations the radiation pressure can be neglected, because the density of radiation is so low.

The hydrodynamic equation of continuity,

$$\frac{D\rho}{Dt} = -\rho \, \nabla \cdot \mathbf{v}, \tag{6.3}$$

also relates the density and velocity fields.

The energy equation is a generalization of the thermal balance equation (3.33),

$$\frac{DU}{Dt} \equiv \frac{D}{Dt}\left(\frac{3}{2}\sum_j N_j kT\right) = G - L + \frac{p}{\rho}\frac{D\rho}{Dt} - U\,\nabla \cdot \mathbf{v}, \tag{6.4}$$

where U is the internal kinetic energy per unit volume, G and L are the energy gain and loss rates per unit volume per unit time discussed in Chapter 3, the next term on the right-hand side of the equation gives the heating rate resulting from compression, and the last term gives the dilation effect, analogous to the term on the right-hand side of equation (6.3). Note that ionization energy is not included on either side of equation (6.4), but the kinetic energy of all particles is, so the sum includes all atoms and ions as well as electrons, and is dominated (in the ionized gas) by N_p, N_e, $N_{\mathrm{He^+}}$, and $N_{\mathrm{He^{++}}}$. It is a reasonably good approximation to assume, as we have in equation (6.4), that all the ionized species are in temperature equilibrium with one another, because the Coulomb-scattering cross sections are so large, and the relaxation times are correspondingly short.

It is somewhat more convenient to rewrite equation (6.4) in a form that includes the internal kinetic energy per unit mass, $E = U/\rho$, for it then becomes

$$\frac{DE}{Dt} = \frac{D}{Dt}\left(\frac{U}{\rho}\right) = \frac{1}{\rho}(G - L) - p\,\frac{D\left(\dfrac{1}{\rho}\right)}{Dt}. \tag{6.5}$$

Finally, the ionization equation is a generalization of equation (2.30),

$$\frac{DN(X^{+i})}{Dt} = -N(X^{+i})\int_{\nu_i}^{\infty}\frac{4\pi J_\nu}{h\nu}\,a_\nu(X^{+i})\,d\nu$$

$$+ N(X^{+i+1})N_e\alpha_A(X^{+i}, T)$$

$$- N(X^{+i})N_e\alpha_A(X^{+i-1}, T)$$

$$+ N(X^{+i-1})\int_{\nu_{i-1}}^{\infty}\frac{4\pi J_\nu}{h\nu}\,a_\nu(X^{+i-1})\,d\nu$$

$$- N(X^{+i})\,\nabla \cdot \mathbf{v}. \tag{6.6}$$

The time-dependent equations are thus nonlinear integrodifferential equations, and are sufficiently complicated so that, to date, only vastly simplified problems have been solved.

In addition to the continuous variations in ρ, $\mathbf{v}$, and so on, implied by equations (6.1)–(6.6), there may also be near discontinuities, or shock and ionization fronts, in nebulae. Let us first consider a shock front, across which ρ, $\mathbf{v}$, and p change discontinuously, but the ionization does not change. Actually, of course, a real shock front is not an infinitely sharp discontinuity, but in many situations the mean-free path for atomic collisions (which gives the relaxation length) is so short in comparison with the dimensions of the flow that ρ, $\mathbf{v}$, and p are nearly discontinuous. For this analysis it is most convenient to use a reference system moving with the shock front, for if the motion is steady, this reference system moves with constant velocity. If we assume a plane-steady shock and denote the physical parameters ahead of and behind the shock by subscripts 0 and 1, respectively, then the momentum and mass-conservation conditions across the front, corresponding to equations (6.1) and (6.3), respectively, are, in this special reference system,

$$p_0 + \rho_0 v_0^2 = p_1 + \rho_1 v_1^2, \tag{6.7}$$

$$\rho_0 v_0 = \rho_1 v_1, \tag{6.8}$$

where the velocity components are in the direction of motion perpendicular to the front.

Furthermore, the energy-conservation condition found by integrating equation (6.5) across the front, using the equation of state, (6.2), is that the gas is compressed adiabatically:

$$p = K\rho^{5/3} = K\rho^\gamma. \tag{6.9}$$

This relation is generally used by substituting it into the equation of motion (6.1), taking the dot product with $\mathbf{v}$, and integrating through the front, giving

$$\frac{1}{2} v_0^2 + \frac{5}{2} \frac{p_0}{\rho_0} = \frac{1}{2} v_1^2 + \frac{5}{2} \frac{p_1}{\rho_1}, \tag{6.10}$$

or, for a general γ,

$$\frac{1}{2} v_0^2 + \frac{\gamma}{\gamma - 1} \frac{p_0}{\rho_0} = \frac{1}{2} v_1^2 + \frac{\gamma}{\gamma - 1} \frac{p_1}{\rho_1}. \tag{6.11}$$

Note that the first term on either side of equation (6.10) represents the flow-kinetic energy per unit mass, and the second term may be broken up into two contributions, $(3/2)p/\rho = (3/2)kT/\mu m_H$, the thermal kinetic energy per unit mass, and p/ρ, the compressional contribution to the energy per unit mass. The more general form, (6.11), includes, in addition, the energy contribution of the internal degrees of freedom of the gas molecules.

Equations (6.7), (6.8), and (6.11) are the familiar Rankine-Hugoniot conditions on the discontinuities at a shock front. However, the physical situation in a gaseous nebula is quite different from that in a laboratory shock tube, and as a result the applicable equations often take a different form. To see this, we estimate the order of magnitudes of the various terms in equation (6.4). From the discussion of Chapter 3, and particularly Figures 3.2 and 3.3, we know that the heating and cooling rates G and L are of order $10^{-24} N_e N_p$ erg cm^{-3} sec^{-1}, and if we consider a "typical" nebula with density $N_e \approx N_p \approx 10^3$ cm^{-3}, intermediate between bright planetaries and bright H II regions, $G \approx L \approx 10^{-18}$ erg cm^{-3} sec^{-1}. At the equilibrium temperature $T \approx 10,000°$K, $U \approx 10^{-9}$ erg cm^{-3}, so typical time scales for heating and cooling by radiative processes are $U/G \approx 10^9$ sec ≈ 30 yr. On the other hand, typical velocities in nebulae are of order of a few times the velocity of sound, at most 30 km sec^{-1}, which corresponds to 10^{-12} pc sec^{-1}. Since the sizes of nebulae are typically in the range 0.1 pc (planetary nebulae) to 10 pc (H II regions), the time scales for appreciable expansion or motion are appreciably longer than 10^9 sec, and the heating and cooling rates due to compression and dilation in equation (6.4) are appreciably smaller than the heating and cooling rates due to radiation. Thus to a first approximation, the temperature in the nebula is fixed by radiative processes, independently of the hydrodynamic conditions, and a shock front in a nebula may be considered isothermal. What happens, of course, is that across the actual shock front equation (6.11) applies, and the temperature is higher behind the front than ahead of it. But in the hot region immediately behind the front, the radiation rate is large and the gas is very rapidly cooled, so that relatively close behind the shock the gas is again at the equilibrium temperature, the same temperature as in the gas just ahead of the shock. The jump conditions (6.6), (6.7), and, instead of (6.11),

$$\frac{p_0}{\rho_0} = \frac{p_1}{\rho_1} = \frac{kT}{\mu m_H},$$
(6.12)

corresponding to $\gamma \to 1$ in (6.11), can therefore be applied between the points just ahead of the shock and the points close behind it. The thickness of this "isothermal shock front" is fixed by the radiation rate and is of order (for the conditions assumed previously) 10^{-3} pc.

Next let us consider an ionization front, across which not only ρ, $\mathbf{v}$, and p, but also the degree of ionization change discontinuously. This is a good approximation at the edge of an ionization-bounded region, because, as we have seen, the ionization decreases very sharply in a distance of the order of the mean free path of an ionizing photon, about 10^{-3} pc for the density

$N_H = 10^3 \, \text{cm}^{-3}$ assumed previously. Across this front, the momentum and mass conservation conditions (6.7) and (6.8) still apply. However, the energy-conservation condition is different from that applying at a shock front, because energy is added to the gas crossing the ionization front. Furthermore, the rate of flow of gas through the ionization front is fixed by the flux of ionizing photons arriving at the front, since each ionizing photon produces one electron-ion pair. Thus equation (6.8) becomes

$$\rho_0 v_0 = \rho_1 v_1 = m_i \phi_i, \tag{6.13}$$

where ϕ_i is the flux of ionizing photons,

$$\phi_i = \int_{\nu_0}^{\infty} \frac{\pi F_\nu}{h\nu} \, d\nu, \tag{6.14}$$

and m_i is the mean mass of the ionized gas per newly created electron-ion pair. Let us write the excess kinetic energy per unit mass transferred to the gas in the ionization process as $q^2/2$, defined by the equation

$$\frac{1}{2} m_i q^2 = \int_{\nu_0}^{\infty} \frac{\pi F_\nu}{h\nu} (h\nu - h\nu_0) \, d\nu. \tag{6.15}$$

The conservation of energy across the ionization front may then be expressed in the form

$$\frac{1}{2} v_0^2 + \frac{5}{2} \frac{p_0}{\rho_0} + \frac{1}{2} q^2 = \frac{1}{2} v_1^2 + \frac{5}{2} \frac{p_1}{\rho_1} \tag{6.16}$$

instead of (6.10), in which the extra term on the left-hand side represents the kinetic energy per unit mass released in the photoionization process.

Once again, however, we note that ordinarily in gaseous nebulae the radiative cooling is quite rapid, and as a result a short distance behind the front the temperature reaches the equilibrium values set by the balance between radiative heating and cooling. This does not, however, lead to an isothermal ionization front, because the ionization conditions and hence the heating and cooling rates are quite different on the two sides of the front. Therefore, instead of equations (6.12) or (6.16) we have the conditions

$$\frac{p_0}{\rho_0} = \frac{kT_0}{\mu_0 m_H}$$

and $\tag{6.17}$

$$\frac{p_1}{\rho_1} = \frac{kT_1}{\mu_1 m_H},$$

where T_0 and T_1 are constants fixed by the heating and cooling rates in the H^0 region ahead of the shock and in the H^+ region behind it, respectively, and μ_0 and μ_1 are the corresponding mean molecular weights. Very rough order-of-magnitude estimates are $T_0 \approx 100\,°K$, $T_1 \approx 10,000\,°K$, $\mu_0 \approx 1$, $\mu_1 \approx 1/2$.

6.3 Ionization Fronts and Expanding H^+ Regions

This section will first consider the shock fronts that can occur in the ionized H^+ regions and in the neutral H^0 regions outside them. The ionization fronts that separate the two regions will then be discussed and classified, and finally this classification will be used to describe the evolution and expansion of the H^+ region formed when a hot star is formed in an initially neutral H^0 region.

The jump conditions (6.7), (6.8), and (6.11) relate ρ_0, v_0, and p_0, the physical conditions ahead of the front, with ρ_1, v_1, and p_1, the corresponding conditions behind the front. These equations may be solved to give any three of these quantities in terms of any other three; for our purposes, it is most convenient to consider ρ_0 and p_0 given, and to express the ratios ρ_1/ρ_0 and p_1/p_0 in terms of the Mach number M of the shock front. This can most conveniently be expressed in terms of c_0, the velocity of sound in the undisturbed region ahead of the shock,

$$c_0 = \sqrt{\frac{\gamma p_0}{\rho_0}} = \sqrt{\frac{\gamma k T_0}{\mu_0 m_H}}. \tag{6.18}$$

For example, for an isothermal ($\gamma = 1$) shock in an H^0 region, with $T = 10^2\,°K$, $c_0 \approx 0.9$ km sec^{-1}, while for an adiabatic ($\gamma = 5/3$) shock in an H^+ region with $T = 10^4\,°K$, $c_0 = 17$ km sec^{-1}. Then the Mach number is defined as

$$M = \frac{|v_0|}{c_0}, \tag{6.19}$$

the ratio of the speed of the shock, with respect to the gas ahead of the front, to the sound speed in this gas. The Mach number ranges between the limits $M \to 1$ for a weak shock, which, in this limit is just an infinitesimal disturbance propagating with the velocity of sound, to $M \to \infty$ for a strong shock propagating extremely supersonically.

It is straightforward to show that, in terms of this parameter, the ratio of pressures behind and ahead of the shock is

$$\frac{p_1}{p_0} = \frac{2\gamma}{\gamma + 1} M^2 - \frac{\gamma - 1}{\gamma + 1}, \tag{6.20}$$

while the ratio of densities is

$$\frac{\rho_1}{\rho_0} = \frac{(\gamma + 1)M^2}{(\gamma - 1)M^2 + 2}. \tag{6.21}$$

Note that for a weak shock, $p_1/p_0 \rightarrow 1$ and $\rho_1/\rho_0 \rightarrow 1$, and that for a strong shock, $p_1/p_0 \rightarrow \infty$, but $\rho_1/\rho_0 \rightarrow (\gamma + 1)/(\gamma - 1) = 4$ for an adiabatic shock, and approaches infinity for an isothermal shock. Thus very great compressions occur behind strong isothermal shocks.

Across an ionization front, the jump conditions described by (6.7), (6.8), and (6.12) or (6.17) relate ρ_0, v_0, and p_0 with ρ_1, v_1, and p_1, while (6.14) gives the flux through the front. Let us consider the simplified form of the conditions described by (6.17) and correspondingly express the results in terms of the isothermal sound speeds with $\gamma = 1$ in equation (6.18). Then, solving for the ratio of densities,

$$\frac{\rho_1}{\rho_0} = \frac{c_0^2 + v_0^2 \pm [(c_0^2 + v_0^2)^2 - 4c_1^2 v_0^2]^{1/2}}{2c_1^2}. \tag{6.22}$$

Physically ρ_1/ρ_0 must be real, and therefore there are two allowed ranges of speed of the ionization front,

$$v_0 \geqslant c_1 + \sqrt{c_1^2 - c_0^2} \equiv v_R \approx 2c_1, \tag{6.23}$$

or

$$v_0 \leqslant c_1 - \sqrt{c_1^2 - c_0^2} \equiv v_D \approx \frac{c_0^2}{2c_1}, \tag{6.24}$$

where the approximations apply with $c_1 \gg c_0$, which, as we have seen previously, is the case in H II regions. The higher critical velocity v_R is the velocity of an "R-critical" front; here R stands for "rare" or "low-density" gas, since, for a fixed ϕ_i as $\rho_0 \rightarrow 0$, $v_0 \rightarrow \infty$ and must ultimately become greater than v_R. Likewise, the lower critical velocity v_D is the velocity of a D-critical front, with D standing for "dense" or "high-density" gas. Ionization fronts with $v_0 > v_R$ are called R-type fronts, and since $c_1 > c_0$, $v_R > c_0$, and these fronts move supersonically into the undisturbed gas ahead of them, while D-type fronts with $v_0 < v_D < c_0$ move subsonically with respect to the gas ahead of them.

Let us consider the evolution of the H⁺ region that would form if a hot star were instantaneously "turned on" in an infinite homogeneous neutral gas cloud. Initially, very close to the star, ϕ_i is large and a (spherical) R-type ionization front moves into the neutral gas. Let us simplify greatly by omitting c_0 and expanding (6.22) for $v_0 \gg c_1$. The results are

$$\frac{\rho_1}{\rho_0} = \begin{cases} \frac{v_0^2}{c_1^2}\left(1 - \frac{c_1^2}{v_0^2}\right) \gg 1 \\ \\ 1 + \frac{c_1^2}{v_0^2} \approx 1 \end{cases} \tag{6.25}$$

for the positive and negative signs, respectively, correct to the second order in c_1/v_0; and these two cases are called strong and weak R-type fronts, respectively. The corresponding velocities from (6.13) are

$$v_1 = \begin{cases} \frac{c_1^2}{v_0} \ll c_1 \\ \\ v_0\left(1 - \frac{c_1^2}{v_0^2}\right) \approx v_0 \gg c_1 \end{cases}, \tag{6.26}$$

respectively. Thus in a strong R-type front, the velocity of the ionized gas behind the front is subsonic with respect to the front, and the density ratio is large; on the other hand, in a weak R-type front, the velocity of the ionized gas behind the front is supersonic, and the density ratio is close to unity. A strong R-type front cannot exist in nature, because disturbances in the ionized gas behind it continually catch up with it and weaken it; the initial growth of the H^+ region occurs as a weak R-type front runs out into the neutral gas, leaving the ionized gas behind it only slightly compressed and moving outward with a subsonic velocity (in a reference system fixed in space)

$$v_0 - v_1 = c_1\left(\frac{c_1}{v_0}\right) \ll c_1. \tag{6.27}$$

Though these analytic results only hold to the first order in c_1/v_0, and the zeroth order in c_0/c_1, the general description is valid so long as the ionization front remains weak R-type.

However, as the front runs out into the neutral gas, the ionizing flux ϕ_i decreases both because of geometrical dilution and because of recombinations and subsequent absorption of ionizing photons interior to the front. Thus, from equation (6.13), v_0 decreases and ultimately reaches v_R, and from this time onward the simple R-type front can no longer exist. At this point, $\rho_1/\rho_0 = 2$ and $v_1 = c_1$ (again to the zeroth order in c_0/c_1), that is, the ionized gas behind the front is moving just sonically with respect to the front. At this moment a shock front breaks off from the ionization front, and the now *D*-critical ionization front follows it into the precompressed neutral gas. As time progresses, the shock front gradually weakens (because of the geometrical divergence) and the ionization front continues as a strong *D*-type front, with a large density jump.

This behavior is shown graphically in Figure 6.1, the result of the numerical integration (with a simplified cooling law) of the system of partial differential equations described previously. The graphs show the velocity and density as functions of radial coordinates at nine consecutive instants of time ranging from 2.2×10^4 yr to 2.0×10^6 yr after turn-on of a model O7 star with initial density 6.5 cm^{-3}. For the first two time steps, the ionization front is weak *R*-type, but between 7.8×10^4 yr and 9.0×10^4 yr, it becomes *R*-critical and a shock front breaks off, which can be seen as the discontinuity just slightly ahead of the ionization front at 9.0×10^4 yr. With advancing time the shock slowly advances with respect to the ionization

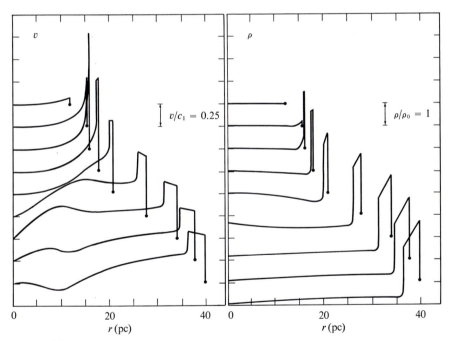

FIGURE 6.1
Simplified model of expanding H II region with initial $N_H = 6.4$ cm^{-3} and $v = 0$, around an O7 star that is turned on at $t = 0$. Left-hand side shows v/c_1 and right-hand side shows ρ/ρ_0, both as functions of r in pc. Successive time steps shown are 2.2×10^4 yr, 7.8×10^4 yr, 9.0×10^4 yr, 1.8×10^5 yr, 3.6×10^5 yr, 9×10^5 yr, 1.4×10^6 yr, 1.8×10^6 yr, and 2×10^6 yr. Each time is displaced downward by $v/c_1 = 0.25$ and by $\rho/\rho_0 = 1$ from the previous time.

front, compressing the neutral gas, while the ionized gas within the H^+ zone expands and the density decreases, so that by 2.0×10^6 yr it is, on the average, only about 0.2 of the density in the undisturbed H^0 zone. If the integration were carried further in time, the shock would weaken still further and the density in the H^+ zone would decrease to the final pressure-equilibrium value

$$\frac{\rho_1}{\rho_0} = \frac{T_0}{2T_1} \approx \frac{1}{2 \times 10^2}, \tag{6.28}$$

but in real nebulae the ionizing star burns out long before this stage is reached.

6.4   Comparisons with Observational Measurements

The comparatively straightforward theory of expanding H II regions given in the previous section is exceedingly difficult to check observationally. The main problem is that there are no nearby nebulae to which it directly applies. As the photographs in this book show, actual nebulae have a very complicated, nonuniform density distribution, directly apparent by the brightness fluctuations that can be seen down to the smallest observable scale. Since these brightness fluctuations are already integrated along the line of sight by the observational technique itself, the actual small-scale density variations must be even more extreme. This is readily confirmed by the measurements of electron density from [O II] and [S II] line ratios, as explained in Section 5.5. Thus the basic picture of a homogeneous "infinite" cloud ionized by a single star within it does not apply except as a very rough approximation. Furthermore, it is quite unlikely that the initial velocity is everywhere zero, as is assumed in the available calculations.

We may nevertheless hope that the calculated results will also be true in some very large-scale average of the velocity field over space. Since the expected velocities are relatively small, of order 10 km sec^{-1} (the isothermal velocity of sound), quite high-dispersion spectral measurements are required, which are difficult to obtain because of the low surface brightness of typical nebulae. Therefore, only a few of the very brightest objects have been studied.

By far the most complete observational study is available for NGC 1976, the Orion Nebula. In the central brightest regions, multislit spectrograms have been obtained covering an area of $4' \times 4'$ centered on θ^1 Ori, the exciting multiple star, at a dispersion of 4.5 A mm^{-1}, corresponding to an instrumental profile with full width at half maximum of approximately 9 km sec^{-1}, and a probable error of measurement of the peak of a line of approximately 1 km sec^{-1}. An example of the [O III] $\lambda5007$ images in one of these spectrograms is shown in Figure 6.2. On these spectrograms [O II] $\lambda3726$,

Hγ and [O III] λ5007 were measured for velocity at a rectangular grid of points separated by 1″.3 in each direction, and in addition the line profiles were measured at selected points. Further from the center of NGC 1976, spectrograms are available at several regions at greater distances (up to about 10′) from θ^1 Ori at the lower dispersion of approximately 9.2 A mm^{-1}. In addition, Fabry-Perot line profiles of [O III] λ5007 with an angular

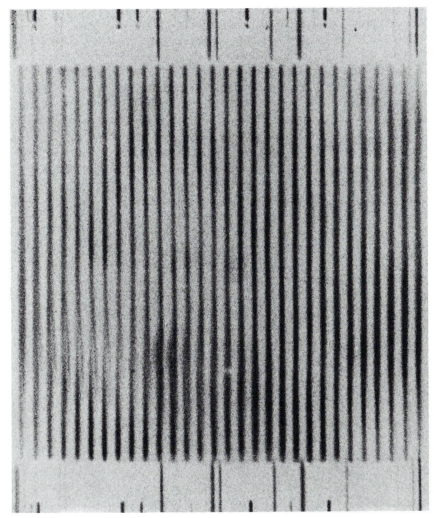

FIGURE 6.2
Multislit image of [O III] λ5007 in an area of NGC 1976, the Orion Nebula, with center 7′.4E, 27″.0 S of θ^1 Ori C. Dimensions are 35″ E-W (along the slits) and 41″ N-S (along dispersion). Slits are separated by 1″.3. (*North is at right, east is at top.*) Note how the broadening, doubling, and intensity change from point to point. (*Hale Observatories photograph.*)

resolution of 6″.5 or 13″ and an instrumental profile with a width of about 4 km sec⁻¹ are available for the central brightest region.

The measured velocities show no very strong evidence of expansion of the nebula similar to that calculated for an initially homogeneous cloud in the previous section. As can be seen in Figure 6.3, there is no clearly apparent pattern of maximum velocity of approach and recession along the line through the center of the nebula, dropping to zero radial velocities at the edges of the nebula. The mean radial velocity of the [O III] projected in the center of the nebula is about 10 km sec⁻¹ more negative than the radial velocity of the exciting stars—the gas is approaching the earth with this velocity relative to the stars, which would indicate expansion if there were no gas behind the stars. However, there is no systematic trend of velocity across the nebula, and furthermore, the [O II] velocity of approach is 3 km sec⁻¹ smaller than the [O III] velocity of approach, though [O II] must be emitted further out in the nebula on the average and should therefore have a larger velocity of approach according to the expansion model.

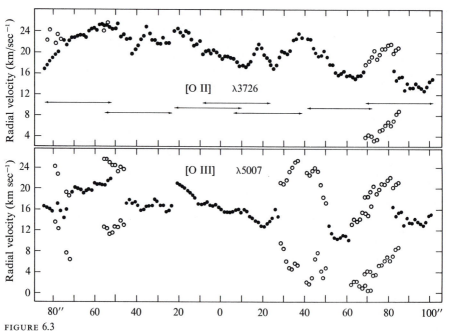

FIGURE 6.3
Radial velocities measured in [O II] and [O III] lines in NGC 1976 along an E-W line 25″ S of θ^1 Ori C. Solid circles indicate points at which line is measured single; open circles indicate points at which line is measured double. The length measured on each separate spectral plate is indicated by an arrow.

The main impression given by the measured velocities in NGC 1976 is one of a turbulent velocity field. First of all, the line profiles at many points in the nebula are approximately Gaussian, but have line widths greater than the thermal width, corresponding quantitatively to a radial velocity dispersion (root-mean-square deviation from the mean) of from 4 km sec^{-1} to 12 km sec^{-1} after the thermal Doppler broadening for $T = 10,000\,°K$ has been removed. This dispersion represents the effect of motions along the line of sight integrated through the whole nebula. At many other points in the nebula, the emission lines are split or double—that is, the profiles are not Gaussian but show double peaks, with separations ranging from 10 km sec^{-1} to 20 km sec^{-1}. Thus, in the regions of line doubling, there are velocity differences along the line of sight through the nebula as large as twice the velocity of sound. The regions in which line doubling occurs are continuous with regions with single lines, and the weighted average velocity of the two components is continuous with the velocity measured from the "single" line just outside these regions. Further, in these regions the resolved [O II] and [O III] components often have different relative intensities, showing that there are differences between the ionization, temperature, or density in the two emitting volumes with different mean velocities. Finally, there are regions where the lines, though not resolved into two components, have strongly non-Gaussian profiles, evidently representing lines of sight with less extreme velocity variations than those that give rise to line splitting.

The measured radial velocities of the peaks of the lines vary with position in the nebula, again with no regular pattern such as would be expected from an expanding sphere, but in an irregular way, with the root-mean-square radial-velocity difference between two positions increasing approximately as the 1/3 power of the distance between them. The observed situation is evidently much closer to what is ordinarily called turbulence than to expansion. There is no obvious correlation between the velocity variations and the apparent surface brightness variations, though the surface brightness variations show that there are tremendous density variations within the nebula.

It seems most likely that the energy source driving the observed turbulence is the primary photoionization process. Dense unionized cold "clouds" probably are contained within the ionized region, and as these regions become ionized, hot gas expands away from them and interacts with gas expanding from other similar clouds. The turbulence thus results from photoionization of an initially very inhomogeneous neutral gas complex. The theoretical description of this situation must be very complicated, and only very small parts of it have yet been worked out—in particular, the ionization of dense spherical neutral "globules" by a spherically symmetric ionizing radiation field. Undoubtedly, future progress must be in the direction of calculations of more realistic models with asymmetric structures, probably including a statistical treatment of the combined effect of many such structures.

In the very recent past, Fabry-Perot interferometers, which have the advantage of giving high wavelength resolution on low-surface-brightness objects, have been developed greatly, and measurements of line profiles in H II regions have become possible. Several other H II regions, such as M 8, M 16, and M 17, have been measured, and the results obtained were rather similar to those obtained with NGC 1976, but the only nebula for which an expansion pattern is clearly seen is NGC 2244, the large Rosette Nebula in Monoceros. A dominantly turbulent velocity field, presumably caused by ionization and expansion of structures within an initially nonhomogeneous H I region, is shown in all H II regions so far observed, and undoubtedly requires very serious theoretical study.

6.5   The Expansion of Planetary Nebulae

The problem of the expansion of planetary nebulae is more satisfactorily understood than the internal velocity distribution of H II regions, apparently largely because the planetary nebulae have more nearly symmetric and initially more nearly homogeneous structures. The earliest high-dispersion spectral studies of the planetary nebulae showed that, in several objects, the emission lines have the double "bowed" appearance shown in Figure 6.4. Later, more complete observational studies of many nebulae showed

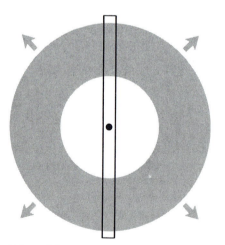

λ ⟶

FIGURE 6.4

Diagram on left shows rectangular slit of spectrograph superimposed on expanding, idealized, spherically symmetric planetary nebula. Resulting image of a spectral line emitted by the planetary, split by twice the velocity of expansion at the center and with splitting decreasing continuously to 0 at edges, is shown on right.

that in nearly all of them the emission lines are double at the center of the nebula; the line splitting is typically of order 50 km sec^{-1} between the two peaks, but decreases continuously to 0 km sec^{-1} at the apparent edge of the nebula. This can be understood on the basis of the approximately radial expansion of the nebula from the central star, with a typical expansion velocity of order 25 km sec^{-1}. Observations further show that there is a systematic variation of expansion velocity with degree of ionization: The ions of highest ionization have the lowest measured expansion velocity, while the ions of lowest ionization have the highest expansion velocity. Since the degree of ionization decreases outward from the star, this observation clearly shows that the expansion velocity increases outward. In this section this expansion will first be discussed theoretically, and then the theory will be compared in detail with the available observational results.

Planetary nebulae are hot ionized gas clouds, which, once formed, must expand into the near vacuum that surrounds them. The expansion of a gas cloud into a vacuum is an old problem that was first treated by Riemann, and only the results will be given here. In the expansion of a finite homogeneous gas cloud, $\rho = \rho_0 = $ constant, released at $t = 0$ from the state of rest $v = 0$, and following the adiabatic equation (6.9), the edge of the gas cloud expands outward with a velocity given by

$$v_e = \frac{2}{\gamma - 1} c_\gamma, \qquad (6.29)$$

while a rarefaction wave moves inward into the undisturbed gas with the adiabatic sound speed c_γ. Thus, at a later time t, the rarefaction wave has reached a radius

$$r_i = r_o - c_\gamma t, \qquad (6.30)$$

where r_o is the initial radius of the gas cloud, while the outer edge has reached a point

$$r_e = r_o + v_e t, \qquad (6.31)$$

and all the gas between these two radii is moving outward with velocity increasing from 0 at r_i to v_e at r_e. For a spherical nebula, the inward-running rarefaction wave ultimately reaches the center of the nebula and is reflected, and the gas near the center is then further accelerated outward.

In an actual planetary nebula, the adiabatic equation (6.9) is not a very good approximation because, as was discussed previously, the heating and cooling are mainly by radiation, and the resulting flow is very nearly isothermal except at extremely low densities. The isothermal approximation corresponds to the limit $\gamma \to 1$ in equations (6.29) to (6.31), in which case the outer edge of the nebula expands with velocity $v_e \to \infty$, but the density within the rarefaction wave falls off exponentially, so the bulk of the gas has a velocity not much higher than the sound velocity.

To go beyond this description, it is necessary to integrate numerically the hydrodynamic equations of Section 6.2. This has been done for a few specific models of planetary nebulae, assuming complete spherical symmetry and an idealized radiative-cooling law, and one set of results is shown in Figure 6.5. Here the initial configuration was a spherical shell, with inner radius 2.4×10^{17} cm and outer radius 3.0×10^{17} cm, set into motion with

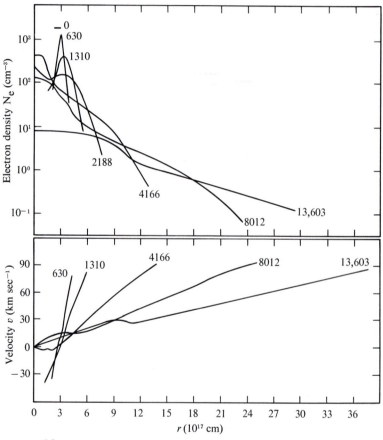

FIGURE 6.5

Top diagram shows calculated variation of electron density N_e with radius r at several times (given, in years, on curves) after expansion begins at $t = 0$ for model planetary nebula described in text. Bottom diagram shows calculated variation of expansion velocity v for same models. Initial homogeneous density distribution $N_e = 1.66 \times 10^3$ cm^{-3} is shown for $t = 0$; initial velocity is $v = 20$ km sec^{-1} at all r.

$v = 20$ km sec^{-1} at $t = 0$; the density and velocity fields are shown at several subsequent times. Notice that the expansion velocity at all times increases more or less linearly outward, a common result in all spherical expansion problems, since the central boundary condition ensures that the $v = 0$ there, and the material at the outer edge generally has the highest expansion velocity. The main difficulty with these models is that the calculated density profiles shown in Figure 6.5 characteristically have the highest density near $r = 0$, while most observed planetaries, including nearly all those for which velocities of expansion have been measured, have a ring-formed appearance in the sky; that is, they appear to be objects with lower densities near their centers. Likewise, Figure 6.5 shows that, at the edge of the calculated model expanding planetary nebula, the density decreases outward with a long tail, which would be observed as a diffuse outer edge, while most real planetaries have a more nearly sharp outer edge. Some astronomers have argued that the forms of many planetary nebulae observed in various projections suggest that they are toroidal objects, rather than spherically symmetric shells. Naturally, the problem of expansion of a toroidal object would be much more difficult to calculate numerically, and in fact at the present time no such calculations are available.

However, if a spherically symmetric model is assumed, then in order to get the observed central "hole" of a typical planetary nebula, it is necessary to suppose that some additional repulsive force is exerted on the nebular gas from the central star, or from the center of the nebula. This is perhaps not completely unrealistic, because there may well be a "stellar wind" of high-speed particles evaporating or flowing from the surface of the central star. A computed model, in which such a stellar wind was taken into account as a pressure exerted on the inner edge of the spherical shell of the planetary nebula, is shown in Figure 6.6. This model shows significantly better agreement with both the shape and the velocity field of a typical planetary nebula, and its success may possibly be taken as a confirmation of the idea of a stellar wind. However, since the pressure taken into account in the calculation results from the stopping of the high-speed "wind" particles in the gas at the inner edge of the planetary nebula, these particles must deliver energy to the gas there, heating it and resulting in additional radiation. No signs of this have yet been observed, but to detect such particles would probably require high-resolution measurements of individual points near the inner edge of the nebula. Also, the calculated temperatures in the outer part of the nebula are quite low, of order 3000°K, because of the expansion cooling. Thus at the present time, the theory of the expansion of planetary nebulae is not in a completely satisfactory state, but theoretical models based on apparently reasonable assumptions roughly fit the available observational data on velocities and densities, and can perhaps be reconciled with the observed line spectra.

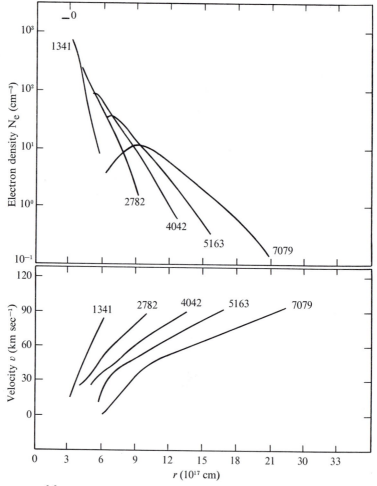

FIGURE 6.6

Top diagram shows calculated variation of electron density N_e with radius r at several times (given, in years, on curves) after expansion begins at $t = 0$ for model planetary nebula with stellar wind described in text. Bottom diagram shows calculated variation of expansion velocity v for same models. Initial conditions are the same as those for models described in Figure 6.5.

Let us turn next to a brief examination of some of the observational data on the expansion velocities of planetary nebulae. Measured velocities (half the separation of the two peaks seen at the center of the nebula) for several ions in a number of fairly typical planetaries are listed in Table 6.1.

In addition, for some of the brighter planetary nebulae, spectrograms taken at a dispersion of about 4.1 A mm^{-1} in the blue and about 6.5 A mm^{-1}

TABLE 6.1
Measured Expansion Velocities in Planetary Nebulae

Nebula	Velocity (km sec^{-1})					
	[O I]	H I	[O II]	[O III]	[Ne III]	[Ne V]
NGC 2392	—	—	53.0	52.6	57.0	0:
NGC 3242	—	20.4	—	19.8	19.5	—
NGC 6210	—	21.0	35.6	21.4	20.8	—
NGC 6572	16.0	—	16.8	—	—	—
NGC 7009	—	21.0	20.4	20.6	19.4	—
NGC 7027	22.8	21.2	23.6	20.4	22.4	19.1
NGC 7662	—	25.8	29.0	26.4	25.9	19.3
IC 418	25.0	17.4	0:	0:	0:	—

in the red, giving line profiles with a velocity resolution of about 5 km sec^{-1}, are available. For instance, the line profiles of NGC 7662 shown in Figure 6.7 result from tracing (at the center of the nebula) the portions of the spectrogram shown in Figure 6.8. The double peaks, with wings extending over a total velocity range of 100 km sec^{-1}, are clearly shown, as well as the fact that the lines are asymmetric, a common feature in planetary nebulae. Though in NGC 7662 the peak with positive radial velocity is stronger, in other nebulae the reverse is true, and this asymmetry is clearly an effect of the departure from complete symmetry of the structure of the nebula itself.

The observed line profile results from the integration along the line of sight of radiation emitted by gas moving with different expansion velocities, further broadened by thermal Doppler motions so that the observed profile $P(V)$ can be represented as an integral

$$P(V) = \text{const} \int_{-\infty}^{\infty} E(U)e^{-m(U-V)^2/2kT} \, dU, \tag{6.32}$$

where $E(V)$ is the distribution function of the emission coefficient in the line per unit radial velocity V for an ion of mass m in an assumed isothermal nebula of temperature T. In the lower part of Figure 6.7, synthetic line profiles calculated from this equation using an approximately triangular distribution function $E(V)$ (half of which is shown in the insert) can be seen to represent the observed profiles quite well. Though other forms of $E(V)$ also fit the observations just as well, all of them require the common feature of a peak at approximately 25 km sec^{-1}, decreasing to nearly zero at approximately 10 km sec^{-1} and 40 km sec^{-1}. The expansion velocity in some nebulae is small enough that the ion with lowest atomic mass, H$^+$, has sufficiently large thermal Doppler broadening so that a double peak is not

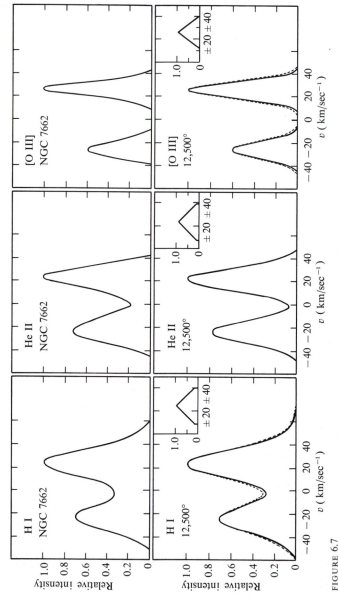

FIGURE 6.7

Top diagrams show observed emission-line profiles at center of planetary nebula NGC 7662. Bottom diagrams show calculated line profiles for simple model of expanding planetary nebula; distribution function of emission coefficient in the line is shown in inserts in upper right-hand corner of each diagram. Dashed lines in H I and [O III] models show model corrected for very slight effects of instrumental broadening.

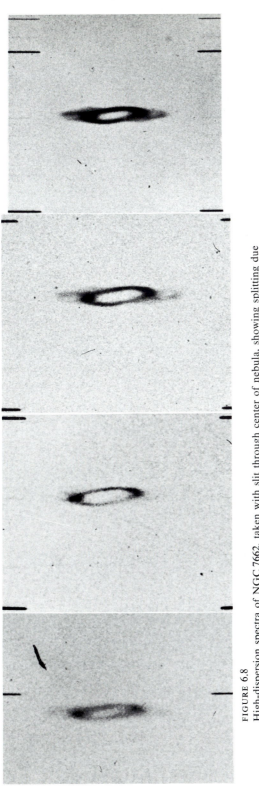

FIGURE 6.8
High-dispersion spectra of NGC 7662, taken with slit through center of nebula, showing splitting due to expansion, as well as line broadening due to expansion and thermal Doppler effect. Hα λ6563 is at left. The remaining photographs show [O III] λ5007 taken with successively longer exposure times (*left to right*). (*Hale Observatories photographs.*)

seen—these are the nebulae measured to have "zero" expansion velocity in Table 6.1. However, in nebulae of this type, for which line profiles have been measured, the same distribution function $E(V)$, which fits the resolved line profile of an ion of higher mass, such as [O III] $\lambda\lambda 4959$, 5007, also fits the unresolved H I profiles.

For some of the nearby planetary nebulae, the average velocity of expansion of 25 km sec^{-1} is large enough that the proper motion of expansion should be marginally detectable with a long-focal-length telescope over a baseline of order 50 yr. Though some measurements have been obtained, to date the results are inconclusive and contradictory. Part of the difficulty is to find sharp, well-defined features, whose positions can be accurately measured. To obtain all results of this type that have been published up to the present time, it was necessary to use first-epoch plates taken on blue-sensitive emulsions, which combine many different emission lines into the image. These plates do not show many sharp features, and it would be a worthwhile long-term program to take new first-epoch plates with a long-focus telescope using red-sensitive emulsions and a filter to restrict the photograph of Hα + [N II] $\lambda\lambda 6548$, 6583, a combination that does show many sharp features.

The main features of the expansion of planetary nebulae seem to be fairly well understood, but a good comparison of Doppler and proper-motion determinations of expansion velocity would give independent data on the distances of the nebulae measured.

References

The importance of hydrodynamical studies of nebulae is obvious, but little serious theoretical work on these problems was done previous to the Symposium on the Motions of Gaseous Masses of Cosmical Dimensions, sponsored by the International Union of Theoretical and Applied Mechanics (IUTAM) and the International Astronomical Union (IAU), held in Paris in 1949, the proceedings of which were published as

> Burgers, J. M., and van de Hulst, H. C., eds. 1951. *Problems of Cosmical Aerodynamics*. Dayton, Ohio: Central Air Documents Office.

This symposium stimulated astronomers, physicists, and gas dynamicists to study nebular problems. Succeeding symposiums have continued to stimulate research in this field:

Burgers, J. M., and van de Hulst, H. C., eds. 1955. *Gas Dynamics of Cosmic Clouds* (IAU Symposium No. 2). Amsterdam: North Holland Publishing Co.

Burgers, J. M., and Thomas, R. N., eds. 1958. *Proceedings of the Third Symposium on Cosmical Gas Dynamics* (IAU Symposium No. 8). *Rev. Mod. Phys.* **30**, 905.

Habing, H. J., ed. 1970. *Interstellar Gas Dynamics* (IAU Symposium No. 39). Dordrecht: D. Reidel.

All these symposium volumes contain excellent reviews and original papers plus references to almost all published works on the subject of hydrodynamics.

A short monograph on the subject is

Kaplan, S. A. 1966. *Interstellar Gas Dynamics,* ed. F. D. Kahn. London: Pergamon Press.

This book, and indeed any text on hydrodynamics, discusses the hydrodynamical equations in Section 6.2. A very useful book for a review of this material is:

Courant, R., and Friedrich, K. O. 1948. *Supersonic Flow and Shock Waves.* New York: Interscience.

The classification of ionization fronts and shock fronts was exhaustively discussed by

Kahn, F. D. 1954. *Bull. Astr. Inst. Netherlands* **12**, 187.

Axford, W. I. 1961. *Phil. Trans. Roy. Soc. Lon. A.* **253**, 301.

Numerical integrations of the dynamical evolution of an H II region have been carried out by

Mathews, W. G. 1965. *Ap. J.* **142**, 1120.

Lasker, B. M. 1966. *Ap. J.* **143**, 700.

(The model shown in Figure 6.1 is taken from the latter reference.) Much of the material on dynamics of H II regions is summarized in

Mathews, W. G., and O'Dell, C. R. 1969. *Ann. Rev. Astr. and Astrophys.* **7**, 67.

Spitzer, L. 1968. *Diffuse Matter in Space.* New York: Wiley, chap. 5.

By far the most complete observational material on dynamics of H II regions is the Coude-spectrograph survey of NGC 1976 by

Wilson, O. C., Münch, G., Flather, E. M., and Coffeen, M. F. 1959. *Ap. J. Supp.* **4**, 199.

The results are discussed by

Münch, G. 1958. *Rev. Mod. Phys.* **30**, 1035.

(Figure 6.3 is taken from this reference.) Perhaps the best Fabry-Perot observational material on internal velocities that has been published to date is included in the following papers:

Smith, M. G., and Weedman, D. W. 1970. *Ap. J.* **160**, 65.

Meaburn, J. 1971. *Ast. and Ap.* **13**, 110.

Dopita, M. A. 1972. *Ast. and Ap.* **17**, 165.

Dopita, M. A., Gibbons, A. H., and Meaburn, J. 1973. *Ast. and Ap.* **22**, 33.

These four papers include measurements of M 8, M 16, M 17, and M 42. The earlier photographic Fabry-Perot work on nebulae is summarized in

Courtes, G., Louise, R., and Monnet, G. 1968. *Ann. d'Ap.* **31**, 493.

The theory of the ionization of a dense globule is worked out by
  Dyson, J. E. 1968. *Astrophys. and Space Sci.* **1,** 388.

The most complete numerical studies of the expansion of planetary nebulae are:
  Mathews, W. G. 1966. *Ap. J.* **143,** 173.
  Hunter, J. H., and Sofia, S. 1971. *M. N. R. A. S.* **154,** 393.
(The calculated models shown in Figures 6.5 and 6.6 are taken from the first of
these references.) A good general reference on gas dynamics is:
  Stanyukovich, K. P. 1960. *Unsteady Motion of Continuous Media,* trans. J. G.
    Adashko, ed. M. Holt. London: Pergamon Press.
This work includes a good discussion of the analytic results that can be obtained
on the expansion of a spherical gas cloud.

The first spectroscopic measurements of line splitting in planetary nebulae were
made by
  Campbell, W. W., and Moore, J. H. 1918. *Pub. Lick Obs.* **13,** 75.
However, they were unable to interpret the observation fully, and subsequently a
very complete high-dispersion survey was made by Wilson:
  Wilson, O. C. 1948. *Ap. J.* **108,** 201.
  Wilson, O. C. 1950. *Ap. J.* **111,** 279.
This survey includes extensive data and discussion in terms of expansion. (Table
6.1 is taken from the latter reference.) A later, very good summary of the radial-
velocity measurements and their significance is:
  Wilson, O. C. 1958. *Rev. Mod. Phys.* **30,** 1025.
Higher-resolution measurements are available in
  Osterbrock, D. E., Miller, J. S., and Weedman, D. W. 1966. *Ap. J.* **145,** 697.
(Figures 6.7 and 6.8 are taken from the preceding reference.) Higher-resolution
measurements are also available in
  Weedman, D. W. 1968. *Ap. J.* **153,** 49.
  Osterbrock, D. E. 1970. *Ap. J.* **159,** 823.

The observations of proper motions of expansion of planetary nebulae are discussed
by
  Liller, M. H., Welther, B. L., and Liller, W. 1966. *Ap. J.* **144,** 280.
  Liller, M. H., and Liller, W. 1968. *Planetary Nebulae* (IAU Symposium No.
    34). Dordrecht: D. Reidel, p. 38.

7

Interstellar Dust

7.1 Introduction

The discussion in the first six chapters of this book has concentrated entirely on the gas within H II regions and planetary nebulae, and in fact these objects usually are called simply gaseous nebulae. However, they really contain dust particles in addition to the gas, and the effects of this dust on the properties of the nebulae are by no means negligible. Therefore, this chapter will discuss the evidence for the existence of dust in nebulae, its effects on the observational data concerning nebulae, and how the measurements can be corrected for these effects. The measurements of the radiation from the dust itself, and the effect of the dust on the structure and radiation of both H II regions and planetary nebulae, are then considered, and the dynamical effects that result from this dust are briefly discussed.

7.2 Interstellar Extinction

The most obvious effect of interstellar dust is its extinction of the light from distant stars and nebulae. This extinction in the ordinary optical region is largely due to scattering, but it is also partly due to absorption. (Nevertheless, the process is very often referred to as interstellar absorption.) It results in the reduction in the amount of light from a source shining through interstellar dust according to the equation

$$I_\lambda = I_{\lambda 0} e^{-\tau_\lambda}, \tag{7.1}$$

where $I_{\lambda 0}$ is the intensity that would be received at the earth in the absence of interstellar extinction along the line of sight, I_λ is the intensity actually observed, and τ_λ is the optical depth at the wavelength observed. This equation also applies to stars, in which we observe the total flux, with πF_λ substituted for I_λ. Note that the equation is correct when radiation is either absorbed or scattered out of the beam, but only if other radiation is not scattered into the beam. This is a good approximation for all stars and for nebulae that do not themselves contain interstellar dust, but it is incorrect if the dust is mixed with the gas in the nebula and scatters nebular light into the observed ray as well as out of it. (This point will be returned to in Section 7.3.) The interstellar extinction is thus specified by the values of τ_λ along the ray to the star or nebula in question.

The interstellar extinction has been derived for many stars by spectrophotometric measurements of pairs of stars selected because they have identical spectral types. The ratio of their brightnesses,

$$\frac{\pi F_\lambda(1)}{\pi F_\lambda(2)} = \frac{\pi F_{0\lambda}(1) \, e^{-\tau_\lambda(1)}}{\pi F_{0\lambda}(2) \, e^{-\tau_\lambda(2)}}$$

$$= \frac{D_2^2}{D_1^2} e^{-[\tau_\lambda(1) - \tau_\lambda(2)]}, \tag{7.2}$$

depends on the ratio of their distances D_2^2/D_1^2 and on the difference in the optical depths along the two rays. Interstellar extinction, of course, increases toward shorter wavelengths (in common terms, it reddens the light from a star), so by comparing a slightly reddened or nonreddened star with a reddened star, it is possible to determine $\tau_\lambda(1) - \tau_\lambda(2) \approx \tau_\lambda(1)$, essentially the interstellar extinction along the path to the more reddened star, except for an additive constant $2 \ln D_2/D_1$ that is independent of wavelength. The constant is not determined, because it depends on the distance of the reddened star, which is generally not independently known. However, for any kind of interstellar dust, or indeed for any kind of particles, $\tau_\lambda \to 0$ as $\lambda \to \infty$, and it is thus possible to determine the constant approximately by making measurements at sufficiently long wavelengths.

Such measurements have been made over the years for many stars, and from them we have a fairly good idea of the interstellar extinction. They show that, to a good approximation, the form of the wavelength dependence of the interstellar extinction is the same for all stars, and only the amount of extinction varies, so that

$$\tau_\lambda = Cf(\lambda), \tag{7.3}$$

where the constant factor C depends on the star, but the function $f(\lambda)$ is the same for all stars. This result implies physically that, to this same first approximation, the dust everywhere in the observed region of interstellar space is similar. Figure 7.1 shows this standard interstellar extinction expressed relative to the extinction at Hβ, and normalized so that $\tau_{H\gamma} - \tau_{H\alpha}$, the difference in optical depths between the two wavelengths, is 0.50. Notice that the extinction is plotted in terms of reciprocal wavelength (proportional to frequency) because it is nearly linear in this variable. Notice also that the extrapolation to $\lambda \to \infty$ establishes the zero point for the extinction shown on the right-hand scale.

Although the standard form of the interstellar extinction is a good first approximation, careful observations of different stars show that there are in fact variations in the wavelength dependence of the interstellar extinction along different light paths in the Galaxy. The most extreme deviations have been measured for the stars of θ^1 Ori, the Trapezium stars exciting NGC 1976, and in Figure 7.2 their average extinction is compared with the

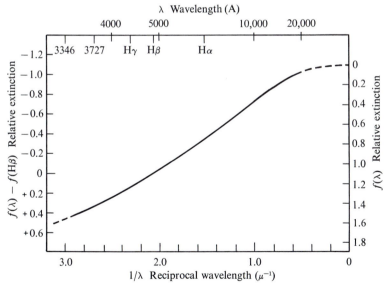

FIGURE 7.1

Standard interstellar extinction curve as a function of wavelength, normalized as described in text and listed in Table 7.1. Note that the left-hand scale gives extinction relative to extinction at Hβ; the right-hand scale shows total extinction and is chosen so that $\tau_\lambda \to 0$ as $\lambda \to \infty$.

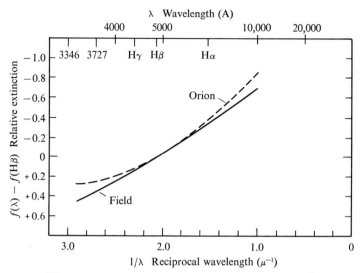

FIGURE 7.2

Average extinction for θ^1 Ori, the Trapezium stars, in NGC 1976, compared with average extinction for stars in Cygnus, Cepheus, Perseus and Monoceros.

average extinction for a large number of stars in Cygnus, Cepheus, Perseus, and Monoceros; the result is essentially the same as the standard extinction curve. It can be seen that the differences are small but not completely insignificant, particularly at the extreme wavelengths shown. Measurements of other stars show that similar deviations from the standard interstellar extinction curve of Figure 7.1 tend to be largest for stars in H II regions. There also seem to be regional variations of the extinction (with galactic longitude), but these variations are even smaller than the differences between the extinction within and without H II regions.

Interstellar extinction naturally makes the observed ratio of intensities of two nebular emission lines $I_{\lambda_1}/I_{\lambda_2}$ differ from their ratio as emitted in the nebula $I_{\lambda_1 0}/I_{\lambda_2 0}$:

$$\frac{I_{\lambda_1}}{I_{\lambda_2}} = \frac{I_{\lambda_1 0}}{I_{\lambda_2 0}} e^{-(\tau_{\lambda_1} - \tau_{\lambda_2})}. \tag{7.4}$$

The observations must be corrected for this effect before they can be discussed physically. As a first approximation, the interstellar extinction can ordinarily be assumed to have the standard form, unless the amount of extinction is very large, so this can also be written

$$\frac{I_{\lambda_1}}{I_{\lambda_2}} = \frac{I_{\lambda_1 0}}{I_{\lambda_2 0}} e^{-C[f(\lambda_1)-f(\lambda_2)]}.$$ (7.5)

Note that only the difference in optical depths at the two wavelengths enters this equation, so the correction depends on the form of the interstellar extinction curve and on the amount of extinction, but not on the more uncertain extrapolation to infinite wavelength. To find the amount of correction, the principle is to use the measured ratio of strengths of two lines for which the relative intensities, as emitted in the nebula, are known independently; thus in equation (7.4) only $\tau_{\lambda_1} - \tau_{\lambda_2}$, or equivalently in equation (7.5) only C, is unknown and can be solved for. Once C is determined, the standard reddening curve (which is listed in Table 7.1) gives the optical depths at all wavelengths.

The ideal line ratio to determine the amount of extinction is one that is completely independent of physical conditions and that is easy to measure in all nebulae. Such an ideal pair of lines does not exist in nature, but various approximations to it do exist and can be used to get a good estimate of the interstellar extinction of a nebula.

The best lines would be a pair with the same upper level, whose intensity ratio would therefore depend only on the ratio of their transition probabilities. An observable case close to this ideal is [S II], in which $^4S-^2P$ $\lambda\lambda4069$, 4076 may be compared with $^2D-^2P$ $\lambda\lambda10287, 10320, 10336, 10370$. Here both multiplets arise from a double upper term rather than from a single level, and the relative populations in the two levels depend slightly on electron

TABLE 7.1
Standard Interstellar Extinction Curve

λ (A)	$1/\lambda$ (μ^{-1})	$f(\lambda) - f(H\beta)$	λ (A)	$1/\lambda$ (μ^{-1})	$f(\lambda) - f(H\beta)$
∞	0.00	-1.09	4545	2.20	$+0.09$
20,000	0.50	-1.02	λ_B	2.30	$+0.15$
12,500	0.80	-0.86	Hγ	2.304	$+0.15$
10,000	1.00	-0.72	4208	2.40	$+0.20$
8333	1.20	-0.56	Hδ	2.438	$+0.22$
7143	1.40	-0.43	4000	2.50	$+0.25$
Hα	1.524	-0.35	3846	2.60	$+0.29$
6256	1.60	-0.29	3727	2.683	$+0.33$
5561	1.80	-0.16	3571	2.80	$+0.39$
λ_V	1.83	-0.14	3346	2.989	$(+0.46)$
5000	2.00	-0.04	3333	3.00	$(+0.46)$
Hβ	2.057	0.00			

density, so the calculated ratio of intensities of the entire multiplets varies between the limits $0.55 \geqslant I(^2D-^2P)/I(^4S-^2P) \geqslant 0.51$ over the range of densities $0 \leqslant N_e \leqslant 10^7 \, \text{cm}^{-3}$. Although this [S II] ratio has been used to determine the interstellar extinction in a few galaxies with emission lines, the planetary nebula NGC 7027 and the supernova remnant NGC 1952, the lines involved are all relatively weak, and in addition the measurement of $I(^2D-^2P)$ is difficult because of contamination due to infrared OH atmospheric emission lines, and also because infrared detectors are relatively insensitive.

A somewhat easier observational method for determining the amount of interstellar extinction of a nebula is to compare an H I Paschen line with a Balmer line from the same upper set of levels. For example, it is possible to compare $P\delta \, \lambda 10049$ with $H\epsilon \, \lambda 3970$, both of which arise from the excited terms with principal quantum number $n = 7$. Since several different upper terms are involved, $7 \, ^2S, 7 \, ^2P, \ldots 7 \, ^2F$, the relative strengths depend slightly on excitation conditions, but as Table 4.4 shows, the variation in Paschen-to-Balmer ratios is quite small over the whole range of temperatures expected in gaseous nebulae. This Paschen-to-Balmer ratio method is, in principle, excellent, but it has the same problems as the [S II] method of contamination by infrared night-sky emission, plus the relative insensitivity of infrared photomultipliers.

Hence the method most frequently used in practice to determine the interstellar extinction is to measure the ratios of two or more H I Balmer lines, for instance, $H\alpha/H\beta$ and $H\beta/H\gamma$. Though the upper levels are not the same for the two lines, the relative insensitivity of the line ratios to temperature, shown in Table 4.2, means that the interstellar extinction can be determined with relatively high precision even though the temperature is only roughly estimated. The Balmer lines are strong and occur in the part of the spectrum that is ordinarily observed, so this method is, at the present time, the one used by far most often. The fact that different pairs of Balmer lines (usually $H\alpha/H\beta$ and $H\beta/H\gamma$) give the same result tends to confirm observationally the recombination theory outlined in Chapter 4.

It can be seen from equation (7.3) that the normalization of the function $f(\lambda)$ that gives the form of the wavelength dependence of the interstellar extinction is arbitrary. The normalization we have adopted is convenient for nebular work, as is the idea of tabulating $f(\lambda) - f(H\beta)$, so that it is simple to correct all emission-line ratios involving $H\beta$, the usual nebular standard reference line. Then, working with logarithms, it is often convenient to write

$$\frac{I_\lambda}{I_{H\beta}} = \frac{I_{\lambda 0}}{I_{H\beta 0}} 10^{-0.434(\tau_\lambda - \tau_{H\beta})}$$

$$= \frac{I_{\lambda 0}}{I_{H\beta 0}} 10^{-c[f(\lambda)-f(H\beta)]} \tag{7.6}$$

and to use $c = 0.434C$ as a measure of the amount of extinction. Nebulae are observed to have a wide range of amount of extinction; for instance, $c \approx 0.02$ for NGC 6720, while the most heavily reddened planetary nebula for which observations have been published to date is probably NGC 7027 with $c = 1.2$.

Naturally, the nebulae with the strongest interstellar extinction are too faint to observe in the optical region, though they can be measured in the radio-frequency region. This suggests still another way to measure the amount of interstellar extinction, namely, to compare the intensity of the radio-frequency continuum at a frequency at which the nebula itself is optically thin to an optical H I recombination line. This is the same principle as that used in comparing two optical lines, except that one of the lines is effectively at infinite wavelength in this method. The intrinsic ratio of intensities $j_\nu/j_{H\beta}$ can be calculated explicitly for any assumed temperature using equations (4.22) and (4.30) and Table 4.4. It depends on the ratio $N_+\langle Z^2\rangle/N_p$, because the free-free emission contains contributions from all ions, but since $N_{He}/N_H \approx 0.10$ and all other elements are smaller, this quantity,

$$\frac{N_+\langle Z^2\rangle}{N_p} \approx 1 + \frac{N_{He^+}}{N_p} + 4\frac{N_{He^{++}}}{N_p}, \tag{7.7}$$

is rather well determined. The temperature dependence of $j_\nu/j_{H\beta}$ is rather low, approximately as $T^{1/3}$, but nevertheless, considerably more rapid than an optical recombination-line ratio such as $H\alpha/H\beta$.

Table 7.2 shows a selection of values of c determined for several planetaries from the most accurate optical and radio measurements, assuming $T \approx 10,000°$K. The probable error of each method is approximately 0.1 in c, and it can be seen that the methods agree fairly well for these accurately measured planetaries. The theory behind these determinations is quite straightforward, and the expected uncertainty because of the range of T in

TABLE 7.2
Interstellar Extinctions for Planetary Nebulae

Nebula	c (Balmer-line method)	c (Radio-frequency-$H\beta$ method)
NGC 6572	0.33	0.24
NGC 6720	0.02	—
NGC 6803	0.59	0.56
NGC 7009	0.10	0.10
NGC 7027	1.15	1.27
IC 418	0.21	0.25

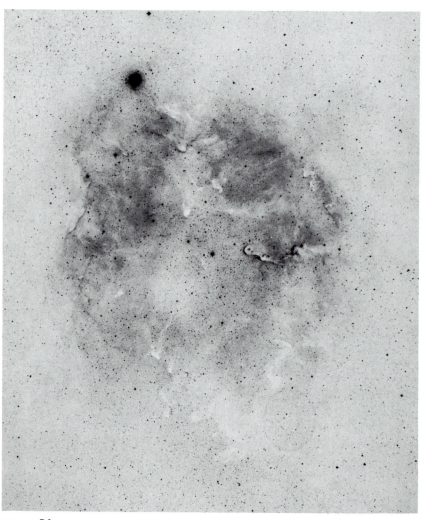

FIGURE 7.3
IC 1396, the H II region in Cepheus, taken with 48-inch Schmidt telescope, red filter and
103a-E plate, emphasizing Hα, [N II]. Note the overlying interstellar extinction, particularly
the large globule (or comet-tail structure), to the east (*right*). (*Hale Observatories
photograph.*)

nebulae is relatively small, so this method, in principle, provides a good absolute determination of the interstellar extinction at the measured optical wavelength; that is, it should determine the extrapolation $\lambda \rightarrow \infty$ of the interstellar extinction curve quite accurately.

7.3 Dust within H II Regions

Dust is certainly present within H II regions, as can clearly be seen on direct photographs. Many nebulae show "absorption" features that cut down the nebular emission and starlight from beyond the nebula. Very dense small

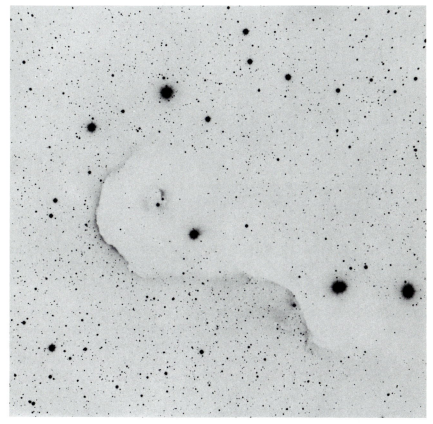

FIGURE 7.4
Large globule (or comet-tail structure) in IC 1396, taken with 200-inch Hale telescope, RG-2 filter and 103a-E plate, emphasizing Hα, [N II]. The very sharp reduction in star density shows that the globule, particularly near its east (*left*) end, is practically opaque. Note the bright edge between the H II region and the globule; this is the ionization front progressing into the dense globule. (*Hale Observatories photograph.*)

features of this kind are often called globules, while others at the edges of nebulae are known as elephant-trunk or comet-tail structures. Many of these absorption features appear to be almost completely dark; this indicates not only that they have a large optical depth at the wavelength of observation (perhaps $\tau \gtrsim 4$ if the surface brightness observed in the globule is a small percentage of that observed just outside it), but also that they are on the

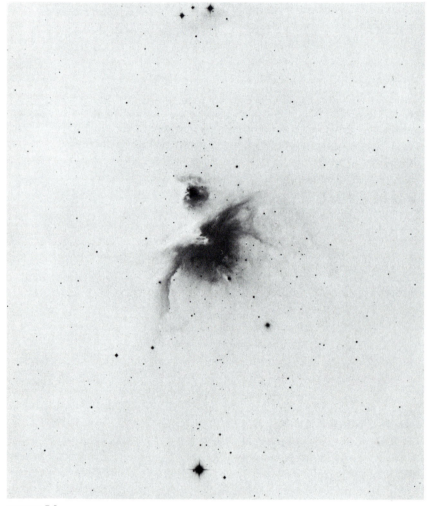

FIGURE 7.5
NGC 1976, the Orion Nebula, taken with 48-inch Schmidt telescope, using Wratten 15 filter and 103a-J plate, in the continuum $\lambda\lambda5100$–5500. Compare this photograph with the Frontispiece (particularly the 40-second exposure), which shows the same nebula in Hα, [N II]. Many differences are apparent. (*Hale Observatories photograph.*)

near side of the nebula, so that very little nebular emission arises between the globule and the observer. A few large-absorption features that are not so close to the near side of the nebula can be seen on photographs; they are features in which the surface brightness is smaller than in the surrounding nebula, but not zero. There must be many more absorption structures with smaller optical depths, or located deeper in the nebula, that are not noticed on ordinary photographs. It is difficult to study these absorption features quantitatively, except to estimate their optical depths, from which the amount of dust can be estimated if its optical properties are known. If, in addition, the gas-to-dust ratio is known, the total mass in the structure can be estimated. We shall return to a consideration of these questions after examining the scattered-light observations of the dust.

The dust particles scatter the continuous radiation of the stars immersed in nebulae, resulting in an observable nebular continuum. Measurements of this continuum must be made with a scanner or interference-filter system designed to avoid the strong nebular-line radiation, and photographs taken in the continuum, such as Figure 7.5, also require filters that avoid strong nebular lines. Measurements of an H recombination line such as Hβ are made at the same time, and from the intensity of that line the expected nebular atomic continuum caused by bound-free and free-free emission can then be calculated using the results of Section 4.3. The atomic contribution is subtracted from the observed continuum, and the remainder, which is considerably larger than the atomic continuum in most observed nebulae, must represent the dust-scattered continuum. This conclusion is directly confirmed by the observation of the He II λ4686 absorption line in the continuous spectrum of one nebula, NGC 1976. This line, of course, cannot arise in absorption in the nebular gas, but is present in the spectrum of the O star in the nebula.

Generally, the observational data cannot be interpreted in a completely straightforward and unique way because of the difficulties caused by complicated (and unknown) geometry and spatial structure of real nebulae. To indicate the principles involved, let us treat the very simplified problem of a spherical, homogeneous nebula illuminated by a single central star. Writing L_ν for the luminosity of the star per unit frequency interval, and further supposing that the nebula is optically thin, the flux of starlight within the nebula at a point distance r from the star is given by

$$\pi F_\nu = \frac{L_\nu}{4\pi r^2}. \tag{7.8}$$

If N_D is the number of dust particles per unit volume in the nebula and Q_λ is their average extinction cross section at the wavelength λ corresponding to the frequency ν, then the extinction per unit volume is $N_D Q_\lambda$, and the emission coefficient per unit volume per unit solid angle due to scattering is then

$$j_\nu = \frac{A_\lambda N_D Q_\lambda \pi F_\nu}{4\pi} = \frac{A_\lambda N_D Q_\lambda L_\nu}{16\pi^2 r^2}, \tag{7.9}$$

where A_λ is the albedo, the fraction of the radiation removed from the flux that is scattered, while $1 - A_\lambda$ is the fraction that is absorbed. Note that in this equation the scattering has been assumed to be spherically symmetric. The intensity of a scattered continuum radiation is then

$$I_\nu(b) = \int j_\nu \, ds$$

$$= \frac{A_\lambda N_D Q_\lambda L_\nu}{8\pi^2} \cdot \frac{1}{b} \cos^{-1} \frac{b}{r_0} \tag{7.10}$$

for a ray with a minimum distance b from the central star in a spherical homogeneous nebula of radius r_0.

This may be compared with the Hβ surface brightness observed from the same nebula, which may, however, be assumed to have a possibly different Strömgren radius r_1 limiting the ionized gas,

$$I_{H\beta}(b) = \int j_{H\beta} \, ds$$

$$= \frac{1}{4\pi} N_p N_e \alpha_{H\beta}^{eff} h\nu_{H\beta} 2 \sqrt{r_1^2 - b^2}. \tag{7.11}$$

In Figure 7.6, these two surface-brightness distributions are compared with observational data for NGC 6514, the most nearly symmetric H II region illuminated by a single dominant central star for which measurements exist. It can be seen that the model is a reasonable representation of this nebula. Then dividing (7.10) and (7.11), the ratio of surface brightness in Hβ to surface brightness in the continuum may be written

$$\frac{I_{H\beta}(b)}{I_\nu(b)} = \left[\frac{N_p N_e \alpha_{H\beta}^{eff} h\nu_{H\beta}}{A_\lambda N_D Q_\lambda} \right]$$

$$\times \left(\frac{4\pi D^2}{L_\nu} \right) \left(\frac{r_0 r_1}{D^2} \right) \left[\frac{\frac{b}{r_0} \sqrt{1 - \frac{b^2}{r_1^2}}}{\cos^{-1} \frac{b}{r_0}} \right]. \tag{7.12}$$

In Equation (7.12) we have inserted D, the distance from the nebula to the observer, and it can be seen that the first factor in square brackets

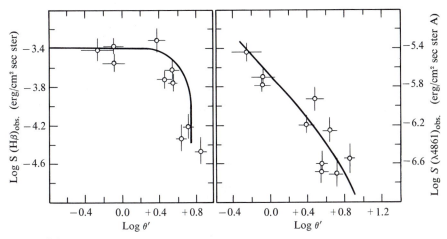

FIGURE 7.6
Diagram on left shows Hβ surface brightness as a function of angular distance from the central star in H II region NGC 6514. Diagram on right shows continuum surface brightness near Hβ (corrected for atomic continuum) as a function of angular distance from the same star.

involves atomic properties and properties of the dust, the second factor is the reciprocal of the flux from the star observed at the earth, the third factor is the product of the angular radii of the nebula in the continuum (r_0/D) and in Hβ (r_1/D), and the fourth factor gives the angular dependence of the surface brightnesses, expressed in dimensionless ratios. Thus the first factor can be determined from measurements of surface brightnesses and of the flux from the star, to the accuracy with which the model fits the data. If the electron density is determined either from the Hβ surface-brightness measurements themselves or from [O II] or [S II] line ratio measurements, the ratio $N_p/A_\lambda N_D Q_\lambda$, proportional to the ratio of densities of gas to dust, is determined; note that this quantity is proportional to the reciprocal of the poorly known electron density. A list of ratios found in this way from continuum observations of several H II regions is given in Table 7.3. For

TABLE 7.3
Gas-to-Dust Ratios in H II Regions

Nebula	Assumed N_e (cm^{-3})	$N_p/A_\lambda N_D Q_{\lambda 4861}$ (cm^{-2})
NGC 1976 (inner)	model	1.4×10^{22}
NGC 1976 (outer)	model	5×10^{20}
NGC 6514	1.3×10^2	4×10^{20}
NGC 6523	4.4×10	2×10^{21}
NGC 6611	5.5×10	2×10^{21}
Field	—	2×10^{21}

NGC 1976, a model in which the density decreases outward, with a range from $N_e \approx 2 \times 10^3$ cm^{-3} in the inner part to $N_e \approx 2 \times 10$ cm^{-3} in the outer part, was used.

The size distribution of the interstellar particles can, of course, be found from the detailed study of their extinction properties. This is a long and complicated subject in itself, approached from measurements of the continuous spectra of stars with different amounts of reddening as discussed in Section 7.2, and it will not be discussed in detail here. It is clear that, for any model of the composition and spectral size of interstellar particles, an extinction curve

$$\tau_\lambda = \langle N_D \rangle Q_\lambda s = C f(\lambda) \tag{7.13}$$

is predicted, and comparison between a predicted curve and observational data, such as Figure 7.1 or 7.2, shows how well the model represents the actual properties of the particles. It is obvious that the larger the wavelength range for which observational data are available, the more specifically the properties of the interstellar particles can be determined. Knowledge of the abundances of the elements in interstellar matter, and physical descriptions of the processes by which the particles are formed and destroyed, also provide information on the properties of the particles.

From all these methods, however, the kinds of dust particles present in interstellar matter are still not completely known, because rather different mixtures of particle distributions can be matched to the same observational data to a fairly good approximation. Until a few years ago, the particles were generally believed to be dielectric, "dirty-ice" particles, consisting mainly of frozen H_2O, CH_4, and NH_3, with smaller amounts of such impurities as Fe and Mg. However, at the time this book is written, measurements of ultraviolet extinction of stars from the OAO satellites are providing new information on the nature of interstellar particles. These measurements show that, in addition to the dielectric particles, there are also other kinds of interstellar particles, most probably graphite particles and silicate particles. In addition, infrared extinction measurements made with high wavelength resolution have failed to show any sign of H_2O (ice) absorption bands by the particles, as would be expected from dirty-ice particles. However, most of the published research on nebulae deals with dielectric particles alone, and this text will necessarily follow these papers; however, it must be realized that many of the conclusions are tentative and may soon be superseded.

A very rough mean value of the radius of the dielectric particles effective in extinction in the optical region is $a \approx 1.5 \times 10^{-5}$ cm. It is clear, of course, that the particles of this size, which is comparable with the wavelength of light, are easiest to observe, and that the information on extremely large particles is rather incomplete. If we now adopt as a very crude mean $A_\lambda \approx 1$, $Q_{\lambda 4861} \approx \pi a^2 \approx 7 \times 10^{-10}$ cm^2, we find that $N_p/A_\lambda N_D Q_{\lambda 4861} = 2 \times 10^{21}$ cm^{-2}, a typical value from Table 7.3, corresponds to $N_p/N_D \approx 1.5 \times 10^{12}$. If we further assume that the density within the dust particles is $\rho \approx 1$ gm cm^{-3}, we find that the ratio of masses $N_p m_H/N_D m_D \approx 1.5 \times 10^2$. The relative gas-to-dust ratio found in this way within H II regions is, in most cases, quite similar to that in an average region of interstellar space, the "field" of Table 7.3; and the ratio of masses is also actually comparable to the ratio of the mass of hydrogen to the mass of heavy elements in typical astronomical objects. Since the bulk of the mass of the dust particles consists of heavy elements, because neither H nor He can be in solid form at the very low density of interstellar space and only compounds like H_2O, NH_3, and CH_4 are expected in the particles, this shows that a fairly large fraction of the heavy-element content of interstellar space is locked up in dust, and that the material in H II regions is not significantly different from the typical interstellar material in this respect. Therefore, the abundance of an element such as O derived from observations of the gas in a nebula is a lower limit to the actual abundance of O, because significant quantities may be present in the form of dust. Note from Table 7.3 that dust is significantly less abundant in the inner part of NGC 1976; as we shall see in the next section, this may result from the presence of the hot stars there.

Of course, the scattering of stellar continuous radiation within the nebula shows that the emission-line radiation emitted by the gas must also be scattered. If the albedo of the dust were $A_\lambda = 1$ at all wavelengths, this scattering would not affect the total emission-line flux from the whole nebula, because every photon generated within it would escape, although the scattering would transfer the apparent source of photons within the nebula. In reality, of course, $A_\lambda \neq 1$ (although it is relatively high), and some emission-line photons are destroyed by dust within the nebula. Therefore, the procedure for correcting observed nebular emission-line intensities for interstellar extinction described in Section 7.2 is not completely correct, because it is based on stellar measurements, in which radiation scattered by dust along the line of sight does not reach the observer. However, numerical calculations of model nebulae using the best available information on the properties of dust show that corrections determined in the way described are very nearly correct and give very nearly the right relative emission-line intensities. The reason is that the wavelength dependence of the extinction, however it occurs, is relatively smooth, so the observational procedure, which amounts to adopting an amount of extinction that correctly fits the observational data to theoretically known relative-line strengths near both ends of the observed wavelength range, cannot be too far off anywhere within that range.

Finally, let us estimate the amount of dust within a globule with radius

0.05 pc that appears quite opaque, so that it has an optical depth $\tau_{H\beta} \geqslant 4$ along its diameter. Many actual examples with similar properties are known to exist in observed H II regions. Supposing that the dust in the globule has the same properties as the dust in the ionized part of the nebula, we easily see that $N_D \geqslant 2 \times 10^{-8} \, \text{cm}^{-3}$; further supposing that the gas-to-dust ratio is the same, $N_H \geqslant 2 \times 10^4 \, \text{cm}^{-3}$. Thus the observed extinction indicates quite high gas densities in globules of this type.

7.4   Infrared Emission

Dust is also observed in H II regions by its infrared thermal emission. The measurements are relatively recent, depending as they do on the development of sensitive infrared detectors and of a whole observational technique to use these detectors effectively. Absorption and emission in the earth's atmosphere become increasingly important at longer wavelengths, but there are windows through which observations can be made from the ground out to just beyond $\lambda \approx 20 \, \mu$. Most of the still longer wavelength measurements have been made from high-altitude balloons or airplanes, though a few observations have been made from mountain-top observatories through partial windows out to 350 μ. In the infrared, subtraction of the sky emission is always very important, and this is accomplished by switching the observing beam back and forth rapidly between the "object" being measured and the nearby "blank sky." This scheme is highly effective for measurements of stars and other objects of small angular size, but it is clear that nebulae with angular sizes comparable with or larger than the angular separation of the object and reference beams are not detected by this method. Most of the observations are taken with rather broad-band filters, though a few high-resolution measurements show that the infrared radiation is essentially continuous, with no sign of emission lines with $\lambda > 10 \, \mu$.

The observations show that, in many H II regions, the infrared radiation is far greater than the free-free and bound-free continuous radiation predicted from the observed Hβ on radio-frequency intensities. This is true for the three bright H II regions that have been studied in detail in the infrared, NGC 1976, NGC 6523, and NGC 6618, as well as for many other H II regions that have been detected in infrared surveys.

Let us first examine the available observational data on NGC 1976, the best studied of the H II regions. There are at least two infrared unresolved "point" sources in this nebula, both of which apparently are highly luminous, heavily reddened stars. In addition, two extended peaks of intensity are measured at 10 μ and 20 μ; one centered approximately on the Trapezium (nearest the stars θ^1 Ori C and D), and the other centered approximately on the brighter infrared point source (often called the Becklin-Neugebauer object) about 1′ northwest of the Trapezium. Both these peaks, or "infrared nebulae," as they have been called (the first is known as the Trapezium nebula, the second is known as the Kleinmann-Low nebula), have angular sizes of order 30″ to 1′, of the same order as the separation of the beams with which they were measured, and are probably only the brightest and smallest regions of a larger complex of infrared emission. The fluxes from each of these peaks range from 10^3 to 10^4 flux units (1 f u = 10^{-26} watts m^{-2} Hz^{-1} = 10^{-23} erg cm^{-2} Hz^{-1} sec^{-1}) from $10\,\mu$ to $20\,\mu$. A measurement at the much longer $\lambda = 350\,\mu$ with an angular resolution of approximately 1′ shows the Kleinmann-Low nebula to have a flux (above the background that is due to the surrounding parts of NGC 1976) of approximately 3000 f u, while the Trapezium nebula is scarcely distinguishable from the background 1′ north or south, which, in turn, is about 1500 f u above the background approximately 2′ east or west. In measurements with considerably larger beams, the flux from NGC 1976 at $100\,\mu$ was measured as 3.5×10^5 f u within a diameter of approximately 12′, and at $400\,\mu$, approximately 3.0×10^5 f u within a diameter of approximately 8′, indicating that a good deal of infrared radiation also is emitted outside the central bright peak at the position of the Kleinmann-Low nebula.

This measured nebular infrared continuous radiation, of order 10^2 to 10^3 as large as the expected free-free and bound-free continua, can only arise by radiation from dust. To a first crude approximation, the dust emits a black-body spectrum, so measurements at two wavelengths approximately determine its temperature. For instance, in the infrared Trapezium nebula, the color temperature determined from the measured fluxes at $11.6\,\mu$ and $20\,\mu$ is $T_c \approx 220°$K; this must approximately represent the temperature of the dust particles. Presumably, they are heated to this temperature by the absorption of ultraviolet and optical radiation from the Trapezium stars, and possibly also from the nearby nebular gas that is ionized by these same stars. Likewise, the dust observed as the Kleinmann-Low nebula is heated by absorption of shorter wavelength radiation probably emitted by or ultimately due to the Becklin-Neugebauer star. However, the measured intensity of the Trapezium nebula at $11.6\,\mu$ is only about $10^{-3}\,B_\nu(T_c)$; this indicates that it has an effective optical depth of only about $\tau_{11.6\mu} \approx 10^{-3}$.

Furthermore, the description and calculations of black-body spectra are somewhat simplified, for measurements with better frequency resolution show that the continuous spectrum does not accurately fit $I_\nu = \text{const } B_\nu(T)$, for any T, but rather has a relatively sharp peak at $\lambda \approx 10\,\mu$, similar to the sharp peak observed in the infrared emission of many cool stars, such as the M2 Ia star μ Cep. In late-type supergiants, this feature is attributed to a maximum in the emission coefficient, perhaps showing that the circumstellar particles are silicates, which should have a band near this position;

the presence of the feature in the nebular infrared spectrum reveals the presence of similar particles in the nebula.

Though the most complete observational data is available for NGC 1976, similar results, though not so detailed, are available for NGC 6523, in which the Hourglass region is a local peak of infrared nebular emission, and for NGC 6618, in which there are two less intense infrared peaks. Evidently, these peaks are regions of high dust density close to high-luminosity stars, which produce the energy that is absorbed and reradiated by the dust.

One interesting result obtained from measurements of a wide band in the far infrared ($\lambda \approx 400 \, \mu$, the band is actually approximately 45–750 μ) is that the measured infrared flux is roughly proportional to the measured radio-frequency flux, as shown in Figure 7.7. Since the radio-frequency flux from a nebula is proportional to the number of recombinations within the nebula, this means that the infrared emission is also roughly proportional to the number of recombinations; that is, to the number of ionizations, or the number of ionizing photons absorbed in the nebula. A plausible inter-

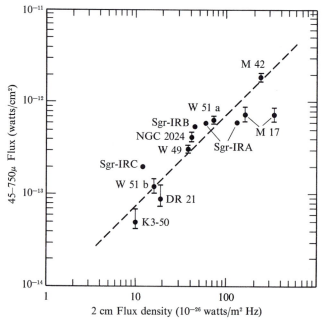

FIGURE 7.7

Measured far infrared (45–750μ) flux and radio-frequency flux for several H II regions, showing proportionality between the two. The dashed line is drawn to fit the data on the average; it corresponds to more infrared emission than can be accounted for from absorption of all the Lα radiation calculated from the radio-frequency emission.

pretation might be that since every ionization by a stellar photon in an optically thick nebula leads ultimately to a recombination and the emission of a $L\alpha$ photon, or of two photons in the $2\,^2S \rightarrow 1\,^2S$ continuum, as explained in Chapters 2 and 4; and since the $L\alpha$ photons are scattered many times by resonance scattering before they can escape, then perhaps every $L\alpha$ photon is absorbed by dust in the nebula and its energy is re-emitted as infrared radiation. According to this interpretation, the ratio of total infrared flux to radio-frequency flux would be

$$\frac{j_{IR}}{j_\nu} = \frac{N_p N_e (\alpha_B - \alpha_{2\,^2S}^{eff}) h\nu_{L\alpha}}{4\pi j_\nu}, \tag{7.14}$$

where the radio-frequency emission coefficient j_ν is given by equations (4.22) and (4.30). α_B, $\alpha_{2\,^2S}^{eff}$, and j_ν depend only weakly on T, and their ratio depends on it even more weakly; j_{IR} and j_ν have the same density dependence, so this ratio is quite well determined.

For $\nu = 1.54 \times 10^{10}$ Hz, the radio frequency used in Figure 7.5, and a representative $T = 7500\,^\circ$K, the calculated ratio from equation (7.14) is $j_{IR}/j_\nu = 1.3 \times 10^{15}$ Hz, but the line drawn through the data corresponds to $j_{IR}/j_\nu = 7.5 \times 10^{15}$ Hz, approximately five times larger. The conclusion is that the infrared emission is larger than can be accounted for by absorption of $L\alpha$ alone; in addition to $L\alpha$, some of the stellar radiation with $\nu < \nu_0$ and possibly also some of the ionizing radiation with $\nu \geqslant \nu_0$ must be absorbed by the dust.

Though the optical continuum measurements showed dust to be present in H II regions, there was no previous observational evidence of dust in planetary nebulae before the infrared measurements were made. However, these measurements show that in many planetaries there is an infrared continuum that is from 10 to 100 times stronger in the $5\,\mu$–$18\,\mu$ region than the extrapolated free-free and bound-free continua. This is shown in Table 7.4, in which the $11\,\mu$ fluxes for some planetaries are compared with the

TABLE 7.4
Infrared and Radio-Frequency Continuum Fluxes of Planetary Nebulae

Nebula	$\pi F_\nu(11\,\mu)$ (f u)	$\pi F_\nu(10$ GHz) (f u)
NGC 6543	54	0.71
NGC 6572	28	1.23
BD $+ 30\,^\circ$ 3639	80	0.52
NGC 7009	10	0.63
NGC 7027	320	6.3
NGC 7662	3	0.54
IC 418	33	1.4

radio-frequency fluxes at 10 GHz, a frequency at which the nebulae are optically thin. Narrow band-width measurements show that most of the infrared radiation has a continuous spectrum, except that the $\lambda 10.52\,\mu$ [S IV] $^2P^0_{1/2}-^2P^0_{3/2}$ emission line has been measured in several planetary nebulae, and the $\lambda 12.8\,\mu$ [Ne II] $^2P^0_{3/2}-^2P^0_{1/2}$ emission line has possibly been detected in IC 418. These emission lines make only a small contribution to the broad-band-measured infrared fluxes, and the fluxes listed in Table 7.4 have been corrected for them and refer to the continuum only.

As in the H II regions, the infrared continuum radiation of planetaries must be due to dust, heated by absorption of stellar radiation and nebular resonance-line radiation, particularly $L\alpha$. In NGC 7027, for which the most detailed infrared spectrum is available, the infrared continuum corresponds approximately to the calculated emission from graphite particles at $T \approx 200°$K; although the composition of the dust is uncertain, the order of magnitude of the temperature is probably correct.

7.5   Survival of Dust Particles in an Ionized Nebula

The observational data obtained from the nebular-scattered optical continuum and thermal infrared continuum show that dust particles exist in H II regions and planetary nebulae. At least in H II regions, their optical properties, and the ratio of amounts of dust to gas, are approximately the same as in the general interstellar medium. Three questions then naturally arise: How are the dust particles initially formed? How long do they survive? How are they ultimately destroyed in nebulae? Much research effort has been expended on these questions, with results that can only very briefly be summarized here.

First let us examine the formation of dust particles. All theoretical and experimental investigations indicate that though dust particles, once formed, can grow by accretion of individual atoms from the interstellar gas, dust particles cannot initially form by atomic collisions at even the highest densities in gaseous nebulae. Thus in planetary nebulae, where the gaseous shell has undoubtedly been ejected by the central star, the dust must have been present in the atmosphere of the star or must have formed during the earliest stages of the process, at the high densities that occurred close to the star. Infrared measurements show that many cool giant and supergiant stars have dust shells around them, so there is observational evidence that this process can occur.

Dust particles in a nebula are immersed in a harsh environment containing both ionized gas, with $T \approx 10,000°$K, and high-energy photons. Collisions of the ions and photons with a dust particle tend to knock atoms or molecules out of its surface and thus tend to destroy it. We must examine very briefly some of the problems connected with the survival of dust particles in a nebula.

First of all, let us consider the electrical charge on a dust grain in a nebula. This charge results from the competition among photoejection of electrons from the solid particle by the ultraviolet photons absorbed by the grain (which tends to make the charge more positive) and captures of positive ions and electrons from the nebular gas (which tend to make the charge more positive and negative, respectively). It is straightforward to write the equilibrium equation for the charge on a grain. The rate of increase of the charge Ze due to photoejection of electrons can be written

$$\left(\frac{dZ}{dt}\right)_{pe} = \pi a^2 \int_{\nu_K}^{\infty} \frac{4\pi J_\nu}{h\nu} \phi_\nu \, d\nu, \tag{7.15}$$

where ϕ_ν is the photodetachment probability ($0 \leqslant \phi_\nu \leqslant 1$) for a photon that strikes the geometrical cross section of the particle. If the dust particle is electrically neutral or has a negative charge, the effective threshold $\nu_K = \nu_c$, the threshold of the material; but if the particle is positively charged, the lowest energy photoelectrons cannot escape, so in general, the threshold is

$$\nu_K = \begin{cases} \nu_c + \dfrac{Ze^2}{ah} & Z > 0 \\ \nu_c & Z \leqslant 0 \end{cases}, \tag{7.16}$$

where $-Ze^2/a$ is the potential energy of an electron at the surface of the particle. The rate of increase of the charge due to capture of electrons is

$$\left(\frac{dZ}{dt}\right)_{ce} = -\pi a^2 N_e \sqrt{\frac{8kT}{\pi m}} \xi_e Y_e, \tag{7.17}$$

where ξ_e is the electron-sticking probability ($0 \leqslant \xi_e \leqslant 1$), and the factor due to the attraction or repulsion of the charge on the particle is

$$Y_e = \begin{cases} 1 + \dfrac{Ze^2}{akT} & Z > 0 \\ e^{Ze^2/akT} & Z \leqslant 0 \end{cases}. \tag{7.18}$$

The rate of increase of the charge caused by capture of protons is, completely analogously,

$$\left(\frac{dZ}{dt}\right)_{cp} = \pi a^2 N_p \sqrt{\frac{8kT}{\pi m_H}} \xi_p Y_p, \tag{7.19}$$

with

$$Y_p = \begin{cases} e^{-Ze^2/akT} & Z > 0 \\ 1 - \dfrac{Ze^2}{akT} & Z \leqslant 0 \end{cases}. \qquad (7.20)$$

Thus the charge on a particle can be found from the solution of the equation

$$\frac{dZ}{dt} = \left(\frac{dZ}{dt}\right)_{pe} + \left(\frac{dZ}{dt}\right)_{ce} + \left(\frac{dZ}{dt}\right)_{cp} = 0, \qquad (7.21)$$

in which the area of the particle cancels out, but the dependence on a through the surface potential remains. Equation (7.21) can be solved numerically for any model of nebula for which the density and the radiation field are known, but the main difficulty is that the properties of the dust particles are only rather poorly known. However, for any apparently reasonable values of the parameters, the general result is that, in the inner part of an ionized nebula, photoejection dominates and the particles are positively charged; while in the outer parts, where the ultraviolet flux is smaller, photoejection is not important and the particles are negatively charged because more electrons, with their higher thermal velocities, strike the particle. As a specific example, for a dirty-ice particle with $a = 3 \times 10^{-5}$ cm, $\xi_e \approx \xi_p \approx 1$, and $\phi_\nu \approx 0.2$ for $h\nu > h\nu_c \approx 12$ eV, the calculated result is that for a representative O5 star with $L \approx 5 \times 10^5 L_\odot$, $T_* \approx 50{,}000°$K, in a nebula with $N_e \approx N_p \approx 16$ cm^{-3}, $Z \approx 380$ at a distance $r = 3.8$ pc from the star, $Z \approx 0$ at $r = 8.5$ pc, and $Z \rightarrow -360$ as $J_\nu \rightarrow 0$.

Knowledge of this charge is important for estimating the rate of sputtering, that is, the knocking out of atoms from the particle by energetic positive ions. The threshold for the process is estimated to be about 2 eV, which is somewhat larger than the mean thermal energy of a proton in a nebula, but of the same order of magnitude as the Coulomb energy at the surface of the particle, $Ze^2/a \approx 0.005\, Z$ eV. Therefore, in the inner part of a nebula, where, say, $Z \approx 400$, the positive charge of the particle raises the threshold significantly, and decreases the sputtering rate, while in the outer part of the nebula, the negative charge of the particle increases the sputtering rate. The efficiency or yield, expressed as the probability that a molecule will be knocked out of the particle per incident-fast proton, is quite uncertain, but according to the best available estimate, $\sim 10^{-3}$, the lifetime of a particle against sputtering is approximately $10^{15}\, a/N_p$ yr in the inner part (with $Z \approx 400$), $2 \times 10^{13}\, a/N_p$ yr where $Z = 0$, and $3 \times 10^{12}\, a/N_p$ yr in the outer part (with $Z \approx -400$). Taking a representative size $a = 3.0 \times 10^{-5}$ cm and $N_p \approx 16$ cm^{-3}, the lifetime of the dust against sputtering is long in comparison with the lifetime of the H II region itself, except possibly in the very outermost parts of the nebula.

Photons tend to destroy a dust particle by heating it to a temperature at which molecules vaporize from the surface. The temperature of the particle, T_D, is given by the equilibrium between the energy absorbed, mostly ultraviolet photons, and the energy emitted, mostly infrared photons:

$$\int_0^\infty 4\pi J_\nu (1 - A_\lambda) Q_\lambda \, d\nu = \int_0^\infty 4\pi B_\nu(T_D)(1 - A_\lambda) Q_\lambda \, d\nu. \qquad (7.22)$$

On the left-hand or absorption side of the equation, since $\lambda \ll a$, the absorption cross section is essentially the same as the geometrical cross section, $(1 - A_\lambda)Q_\lambda \approx \pi a^2$; but on the right-hand or emission side, where $\lambda \gg a$, $(1 - A_\lambda)Q_\lambda \approx \pi a^2 \epsilon (2\pi a/\lambda)$ with $\epsilon \approx 0.1$ for dielectric particles containing a small contamination of such elements as Fe and Mg. As a result,

$$T_D \propto \left(\frac{L}{4\pi r^2 a} \right)^{1/5}, \qquad (7.23)$$

and for a representative particle with $a = 3.0 \times 10^{-5}$ cm, $T_D \approx 100°$K at $r = 3$ pc from the star. The evaporation temperatures of the main constituents of a dielectric particle are $T_v \approx 20°$ for CH_4, $T_v \approx 60°$ for NH_3, and $T_v \approx 100°$ for H_2O, suggesting that CH_4 cannot be held by dust particles anywhere in the nebula, that NH_3 vaporizes except in the outer parts, and that H_2O evaporates only in the innermost parts.

Another possibly important destructive mechanism for dust particles is "optical erosion," which might better be called photon sputtering—the knocking out of atoms or molecules immediately following absorption of a photon. The process can occur either as a result of photoionization of a bonding electron of a molecule in the surface of the particle, or as a result of excitation of a bonding electron to an antibonding level. Very little experimental data is available on this process, and the calculations depend on many unchecked assumptions, but they seem to show that dirty-ice particles would be destroyed by this process in a typical H II region in a relatively short time, of order 10^2 to 10^3 yr, and therefore should not exist at all in nebulae. This is obviously an important question, the answer to which would repay further study.

The overall conclusion is that, for classical dielectric particles, sputtering is not an important mechanism except possibly in the outer parts of a nebula; vaporization is important in the inner parts; and optical erosion may be very important but is only poorly understood. However, the recent ultraviolet and infrared observations show that these particles are by no means as dominant as the earlier optical observations had seemed to indicate, and it is probable that graphite and silicate particles are quite important. Little published work is available on the survival of such particles in ionized nebulae, but in a general way they are more tightly bound and more likely to survive than dirty-ice particles. This conclusion, of course, agrees with the observational certainty that dust particles do exist in H II regions and planetary nebulae.

7.6   Dynamical Effects of Dust in Nebulae

Dust particles in a nebula are subjected to radiation pressure from the central star. However, the coupling between the dust and gas is very strong, so the dust particles do not move through the gas to any appreciable extent, but rather transmit the central repulsive force of radiation pressure to the entire nebula. Let us look at this a little more quantitatively. The radiation force on a dust particle is

$$F_{\text{rad}} = \pi a^2 \int_0^\infty \frac{\pi F_\nu}{c} P_\nu \, d\nu$$

$$= \pi a^2 \int_0^\infty \frac{L_\nu}{4\pi r^2 c} P_\nu \, d\nu \approx \frac{a^2 L}{4r^2 c}, \tag{7.24}$$

where P_ν is the efficiency of the particle for radiation pressure. Since most of the radiation from the hot stars in the nebula has $\lambda \ll a$, $P_\nu \approx 1$ for classical dirty-ice particles. Note, however, that this is not true for very small graphite or silicate particles that may actually exist in the nebula. Notice also that only the radiation force of the central star has been taken into account; the diffuse radiation field is more nearly isotropic and can, to a first approximation, be neglected in considering the motion due to radiation pressure. The force tends to accelerate the particle through the gas, but its velocity is limited by the drag on the particle produced by its interaction with the nebular gas. If the particle is electrically neutral, this drag results from direct collisions of the ions with the grain, and the resulting force is

$$F_{\text{coll}} = \frac{4}{3} N_p \pi a^2 \left(\frac{8kT m_{\text{H}}}{\pi} \right)^{1/2} w, \tag{7.25}$$

where w is the velocity of the particle relative to the gas, assumed to be small in comparison with the mean thermal velocity. Thus the particle is accelerated until the two forces are equal, and reaches a terminal velocity

$$w_t = \frac{3L}{16\pi r^2 c N_p} \left(\frac{\pi}{8kT m_{\text{H}}} \right)^{1/2}, \tag{7.26}$$

which is independent of the particle size. As an example, for a particle at a distance of 3.3 pc from the O star we have been considering, $w_t = 10$ km

$\sec^{-1}$, and the time required for a relative motion of 1 pc with respect to the surrounding gas is about 10^5 yr.

However, for charged particles, the Coulomb force increases the interaction between the positive ions and the particle significantly, and the drag on a charged particle has an additional term,

$$F_{\text{Coul}} \approx \frac{2N_p Z^2 m_{\text{H}}}{T^{3/2}} w, \tag{7.27}$$

with T expressed in °K. Comparison of equation (7.27) with equation (7.25) shows that Coulomb effects dominate if $|Z| \gtrsim 50$, and since, in most regions of the nebula, the particles have a charge greater than this, the terminal velocity is even smaller and the motion of the particle with respect to the gas is smaller yet. Under these conditions the dust particles are essentially frozen to the gas, the radiation pressure on the particles is communicated to the nebular material, and the equation of motion therefore contains an extra term on the right-hand side, so that equation (6.1) becomes

$$\rho \frac{D\mathbf{v}}{Dt} = -\nabla p - \rho \nabla \phi + N_D \frac{a^2 L}{4r^2 c} \mathbf{e}_r. \tag{7.28}$$

Substitution of typical values, including the observationally determined gas-to-dust ratio, shows that the accelerations produced can be appreciable, and the radiation-pressure effects should therefore be taken into account in the calculation of a model of an evolving H II region. An approximate calculation of this type has shown that, with reasonable amounts of dust, old nebulae will tend to develop a central "hole" that has been swept clear of gas by the radiation pressure transmitted through the dust. An example of a real nebula to which this model may apply is NGC 2244, shown in Figure 7.8.

The observational data clearly show that dust does exist in nebulae, but unfortunately its optical and physical properties are not accurately known, so the specific calculations that have been carried out to date must be considered schematic and indicative rather than definitive. Progress in this field may be expected to be rapid as the nature of the particles is determined from the ultraviolet satellite observations.

References

Interstellar extinction is a subject with a long history, most of which is not directly related to the study of gaseous nebulae. Several useful summaries of the available data are:

Hulst, H. C. van de. 1957. *Light Scattering by Small Particles.* New York: Wiley. See especially chap. 21.

Johnson, H. L. 1968. *Nebulae and Interstellar Matter,* ed. B. M. Middlehurst and L. H. Aller. Chicago: University of Chicago Press, chap. 5.

Greenberg, J. M. 1968. *Nebulae and Interstellar Matter,* ed. B. M. Middlehurst and L. H. Aller. Chicago: University of Chicago Press, chap. 6.

Lynds, B. T., and Wickramasinghe, N. C. 1968. *Ann. Rev. Astr. and Astrophys.* **6**, 215.

Whitford, A. E. 1958. *A. J.* **63**, 201.

(The standard interstellar extinction curve of Figure 7.1 and Table 7.1, often called the Whitford interstellar extinction curve, is taken from this last reference, which summarizes a large amount of observational work.)

Some more recent observational results are contained in

Whiteoak, J. B. 1966. *Ap. J.* **144**, 305.

(The preceding reference discusses regional differences in the extinction, and the data for Figure 7.2 are taken from it.)

The differences in extinction between stars in H II regions and stars outside H II regions are carefully studied in the following reference:

Anderson, C. M. 1970. *Ap. J.* **160**, 507.

The [S II] method of measuring the extinction of the light of a nebula is described by

Miller, J. S. 1968, *Ap. J.* **154**, L57.

Applications of the method to gaseous nebulae are contained in

Miller, J. S. 1973. *Ap. J.* **180**, L83.

Schwartz, R. D., and Peimbert, M. 1973. *Ap. Letters* **13**, 157.

The Paschen/Balmer and Balmer-line ratio methods are described parenthetically in many references chiefly devoted to observational data on planetary nebulae. For instance:

Aller, L. H., Bowen, I. S., and Minkowski, R. 1955. *Ap. J.* **122**, 62.

Osterbrock, D. E., Capriotti, E. R., and Bautz, L. P. 1963. *Ap. J.* **138**, 62.

O'Dell, C. R. 1963. *Ap. J.* **138**, 1018.

Miller, J. S. and Mathews, W. G. 1972. *Ap. J.* **172**, 593.

This last reference contains the most accurate measurements of NGC 7027, and shows that calculated Balmer-line and Balmer-continuum ratios, modified by interstellar extinction, agree accurately with the measurements. These authors give convenient interpolation formulas for fitting the standard interstellar extinction curve.

The Balmer radio-frequency method of determining the extinction is described in

Osterbrock, D. E. and Stockhausen, R. E. 1961. *Ap. J.* **133**, 2.

Osterbrock, D. E. 1964. *Ann. Rev. Astr. and Astrophys.* **2**, 95.

Pipher, J. L. and Terzian, Y. 1969. *Ap. J.* **155**, 475.

The most accurate observational results of this method are contained in

Peimbert, M., and Torres-Peimbert, S. 1971. *Bol. Obs. Tonantzintla Tacubaya* **6**, 21.

The essential data of Table 7.2 are taken from this reference, although the numerical results in the table are slightly different because these authors used a different interstellar extinction curve.

The presence of a nebular continuum in diffuse nebulae, resulting from dust scattering of stellar radiation, is an old concept. The fact that it is the dominant contributor to the observed continuum was first quantitatively proved by

Wurm, K., and Rosino, L. 1956. *Mitteilungen Hamburg Sternwarte Bergedorf* **10**, Nr. 103.

They compared photographs of NGC 1976 (similar to the Frontispiece) taken with narrow-band filters, which isolated individual spectral lines with photos of a region in the continuum free of emission lines (similar to Figure 7.5), and showed that the appearance of the nebula in the continuum is different from its appearance in any spectral lines, and that the continuum cannot have an atomic origin and therefore (by implication) must arise from dust. The He II λ4686 absorption line in the continuum of NGC 1976 was observed by

Peimbert, M., and Goldsmith, D. W. 1972. *Ast. and Ap.* **19**, 398.

Quantitative measurements of the continuum surface brightnesses in several nebulae, and comparisons with Hβ surface brightnesses, were made by

O'Dell, C. R., and Hubbard, W. B. 1965. *Ap. J.* **142**, 591.

O'Dell, C. R., Hubbard, W. B., and Peimbert, M. 1966. *Ap. J.* **143**, 743.

The analysis given in the text is based on the second of these references. (Figure 7.4 and Table 7.3 are taken from it.) The simplifying assumptions made in the analysis are not necessary, and more realistic models have been calculated by

Mathis, J. S. 1972. *Ap. J.* **176**, 651.

The classical optical work on interstellar dust, its extinction, and its scattering properties, are very well summarized in the book by van de Hulst mentioned at the beginning of these references. The deduced properties of dielectric interstellar particles, including numerical values for Q_λ, A_λ, and a used in the text, are suggested by this reference. The infrared observations, which set an upper limit to the amount of H_2O in the particles that is considerably lower than would be expected from the optical measurements, are described in

Danielson, R. E., Woolf, N. J., and Gaustad, J. E. 1965. *Ap. J.* **141**, 116.

Knacke, R. F., Cudaback, D. D., and Gaustad, J. E. 1969. *Ap. J.* **158**, 151.

The strong deviation of the observed interstellar extinction in the ultraviolet from the prediction of the dirty-ice particle model was first observed with rocket-borne telescopes by

Stecher, T. P. 1965. *Ap. J.* **142**, 1683.

Stecher, T. P. 1969. *Ap. J.* **157**, L125.

The more recent satellite ultraviolet observations of interstellar extinction are very well summarized, discussed, and analyzed by

Bless, R. C., and Code, A. D. 1972. *Ann. Rev. Astr. and Astrophys.* **10**, 197.

Bless, R. C., and Savage, B. D. 1972. *Ap. J.* **171**, 293.

Gilra, D. P. 1971. *Nature* **229**, 237.

However, further observation, in progress as this book is written, may be expected to improve our knowledge of the properties of interstellar dust.

The effect of nebular scattering on the correction of measured line-intensity ratios for extinction is discussed by

Mathis, J. S. 1970. *Ap. J.* **159**, 263.

Globules were named and discussed by

Bok, B. J., and Reilly, E. F. 1947. *Ap. J.* **105**, 255.

The more recent observational work is summarized in

Bok, B. J., Cordwell, C. S., and Cromwell, R. H. 1971. *Dark Nebulae, Globules and Protostars,* ed. B. T. Lynds. Tucson: University of Arizona Press, p. 33.

Infrared measurements of nebulae have only been made quite recently, so the available information is rapidly growing, and many numerical results given in this section will probably soon be out of date. A good review of the earlier infrared work on nebulae (including planetaries) is

Neugebauer, G., Becklin, E., and Hyland, A. R. 1971. *Ann. Rev. Astr. and Astrophys.* **9**, 67.

However, a considerable number of additional observational results are now available. The most detailed infrared maps of NGC 1976 at 11.6 μ and 20 μ are due to

Ney, E. P., and Allen, D. A. 1969. *Ap. J.* **155**, L193.

The numerical values for these wavelengths quoted in the text are taken from this reference. Measurements further in the infrared region are continued in

Low, F. J., and Aumann, H. H. 1970. *Ap. J.* **162**, L79.

Harper, D. A., and Low, F. J. 1971. *Ap. J.* **165**, L9.

Harper, D. A., Low, F. J., Rieke, G., and Armstrong, K. R. 1972. *Ap. J.* **177**, L21.

The quoted values of the fluxes for the entire nebula (large beam) from an airplane and the individual peaks (small beam) from the ground are taken from the last two of these references. The peak in the spectrum of NGC 1976 near 10 μ, identified with silicates, was observed by

Stein, W. A., and Gillett, F. C. 1969. *Ap. J.* **155**, L197.

Detailed infrared measurements of NGC 6618 and NGC 6523, respectively, are given by

Lemke, D., and Low, F. J. 1972. *Ap. J.* **177**, L53.

Woolf, N. J., Stein, W. A., Gillett, F. C., Merrill, K. M., Becklin, E. E., Neugebauer, G., and Pepin, T. J. 1973. *Ap. J.* **179**, L111.

A survey of a large number of infrared sources, including many H II regions, is included in

Hoffman, W. L., Frederick, C. L., and Emery, R. J. 1971. *Ap. J.* **170**, L89.
The rough proportionality of the wide-band infrared fluxes of H II regions to their free-free radio-frequency fluxes was discovered by

Harper, D. A., and Low, F. J. 1971. *Ap. J.* **165**, L9.
(Figure 7.5 is taken from this reference.)

Key references on infrared radiation from planetary nebulae are:
  Gillett, F. C., Low, F. J., and Stein, W. A. 1967. *Ap. J.* **149**, L97.
  Gillett, F. C., Merrill, K. M., and Stein, W. A. 1972. *Ap. J.* **172**, 367.
  Krishna Swamy, K. S., and O'Dell, C. R. 1968. *Ap. J.* **151**, L61.
  Gillett, F. C., and Stein, W. A. 1969. *Ap. J.* **155**, L97.
  Rank, D. M., Holtz, J. Z., Geballe, T. R., and Townes, C. H. 1970. *Ap. J.* **161**, L185.
  Holtz, J. Z., Geballe, T. R., and Rank, D. M. 1971. *Ap. J.* **164**, L29.
The first reference reports the discovery of an infrared excess in NGC 7027; the second includes the most recent and most complete list of measured fluxes of the 11μ continuum and also of the [S IV] $\lambda 10.52 \mu$ line. In the third reference, the interpretation of the infrared continuum in terms of dust emission is discussed, and the temperature $T \simeq 200°K$ of the particles is estimated. The [Ne II] line is reported to be measured in IC 418 in the fourth reference, and the [S IV] line is reported to be measured in NGC 7027 in the fifth reference; the last reference contains the most complete infrared-line measurements in planetary nebulae published to date, including fluxes of [S IV] $\lambda 10.52 \mu$ in several, as well as upper limits of [Ne II] $\lambda 12.8 \mu$ in many nebulae, including IC 418. However, because of differences in beam size, this measurement may possibly be reconciled with the measurement of this line by Gillett and Stein.

Problems of survival of dust particles in ionized nebulae are treated by
  Mathews, W. G. 1967. *Ap. J.* **147**, 965.
The electrical charges of the particles and also the temperatures they reach as a result of absorption and emission of radiation are discussed in this reference and in

  Spitzer, L. 1968. *Diffuse Matter in Space*. New York: Interscience, chap. 4.
  Flower, D. R. 1972. *Mem. Société Roy. Sci. Liège* **5**, 165.

The sputtering rate for dirty-ice particles has been worked out by
  Barlow, M. J. 1971. *Nature Phys. Sci.* **232**, 152.

The radiation forces on dust particles in H II regions were investigated by
  O'Dell, C. R., and Hubbard, W. B. 1965. *Ap. J.* **143**, 743.
This subject was also explored by Mathews and by Spitzer (chapter 5) in the references just given. The expressions for the drag force on a particle caused by collisions with gas atoms are credited to Epstein and Spitzer:
  Epstein, P. S. 1924. *Phys. Rev.* **23**, 710.

(We have set $\kappa \approx 1$.)

Spitzer, L. 1956. *Physics of Fully Ionized Gases*. New York: Interscience, chap. 5.

(We have set $\ln \Lambda \approx 22.8$.)

The model expanding nebula (taking radiation pressure into account), which ultimately develops a central cavity, was calculated by Mathews in the reference given previously in this section.

8

H II *Regions in the Galactic Context*

8.1   Introduction

In the first five chapters of this book, we examined the equilibrium processes in gaseous nebulae and compared the models calculated on the basis of these ideas with observed H II regions and planetary nebulae. In Chapters 6 and 7 the basic ideas of the internal dynamics of nebulae, and of the properties and consequences of the interstellar dust in nebulae, were discussed, worked out, and compared with observational data. Thus we have a fairly good basis for understanding most of the properties of the nebulae themselves. In this chapter the H II regions are considered in the wider context of galaxies. The discussion includes the distributions of these regions, both in other galaxies and in our own Galaxy so far as they are known, and the regions' galactic kinematics. Then the stars in H II regions and what is known about star formation and H II region formation are examined, and the evolution of H II regions is sketched on the basis of the ideas presented in the earlier chapters. The chapter concludes with a brief discussion of the molecules in H II regions, a very new and important topic that will undoubtedly lead to new understanding of nebulae in the next few years.

8.2   Distribution of H II Regions in Other Galaxies

H II regions can be recognized on direct photographs of other galaxies taken in the radiation of strong nebular emission lines. The best spectral region for this purpose is the red, centered around Hα λ6563 and [N II] λλ6583, 6548. Most of the pictures of nebulae in this and other astronomical books were taken in this way, using various red filters—often red Plexiglass (λ > 6000), which was used in the National Geographic Society-Palomar Observatory Sky Survey, or Schott RG 2 glass (λ > 6300), often used with large telescopes—together with 103a-E plates (λ < 6700). It is possible to use quite narrow-band interference filters for the maximum rejection of unwanted continuum radiation, and comparison of a narrow-band photograph in the nearby continuum permits nearly complete discrimination between H II regions and continuum sources, which are apparently mostly luminous stars and star clusters. A photographic subtraction process, combining a negative of one plate with a positive of the other, can be used to compare the two exposures, though difficulties are caused by the nonlinearity of the photographic process.

Many external galaxies have been surveyed photographically for H II regions in these ways (for example, NGC 628, which is shown in Figure 8.1). In such studies the entire galaxy can be observed (except for the effects of interstellar extinction), and all parts of it are very nearly the same distance from the observer, in contrast to the situation in our own Galaxy where the more distant parts are nearly completely inaccessible to optical observation. These photographic surveys show that essentially all the nearby, well-studied spiral and irregular galaxies contain many H II regions. On the other hand, elliptical and SO galaxies typically do not contain H II regions, although some of them have ionized gas clouds that are often described as giant H II regions at their nuclei.

In spiral galaxies the H II regions are strikingly concentrated along the spiral arms, and in fact are the main objects seen defining the spiral arms in many of the published photographs of galaxies. Often there are no H II regions in the inner parts of the spiral galaxies, but the spiral arms can be seen as concentrated regions of interstellar extinction—"dust" in the terminology used in galactic structure. Evidently, in these regions there is interstellar matter but there are no O stars to ionize it and make it observable as H II regions. Different galaxies have different amounts of dust and different densities of H II regions along the spiral arms, but the concentration of H II regions along relatively narrow spiral arms and spurs is a general feature of spiral galaxies.

In irregular galaxies the distribution of H II regions is less well organized. In some of the galaxies classified as irregular, features resembling spiral arms can be traced in the distribution of H II regions, but in other irregular

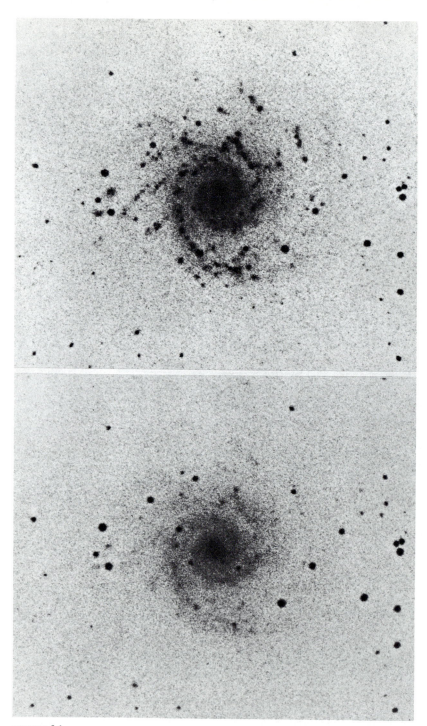

FIGURE 8.1

NGC 628, a nearly face-on spiral galaxy. Top photograph, taken using deep-red filter,
emphasizes H II region; bottom photograph, taken using yellow-orange filter, suppresses
emission nebulae. Exposure times were chosen so that stars appear similar in two
photographs. Note how the H II regions fall along the spiral arms in the top
photograph. Original plates were taken with Yerkes 40-inch refractor and f/2 reducing
camera. (*Yerkes Observatory photographs.*)

galaxies the distribution of H II regions is often far less symmetric; one or more areas may contain many H II regions, but other areas may be essentially devoid of H II regions.

From the direct photographs, it is clear that spiral galaxies are highly flattened, plane systems, and that the H II regions are strongly concentrated, not only to the spiral arms but to the galactic plane in which the arms lie. Photographs or maps from which the effects of the inclination of the plane of the galaxy to the line of sight have been removed by a linear transformation have very symmetric structures that can only result if the H II regions are nearly in a plane.

Because the light of H II regions is largely concentrated into a relatively few spectral lines, they are favorite objects for spectroscopic determinations of radial velocities in external galaxies. For some of the larger nearby galaxies, extensive lists of radial velocities of H II regions are available, chiefly from Hα and [N II]; but in earlier work this information was obtained from [O II] $\lambda3727$, Hβ and [O III] $\lambda\lambda4959, 5007$. These measurements show that, at a particular point in a galaxy, the dispersion of the radial velocities of the H II regions is quite small, and that, when corrected for effects of projection, the velocities show the familiar galactic rotation pattern. Indeed, much of the available information on galactic rotation and the masses of spiral galaxies has been derived from these radial velocity measurements of H II regions.

The spectra of some of the brighter H II regions in external galaxies have been studied and are quite similar to the spectra of H II regions in our Galaxy. Some of the best quantitative information on helium abundances in other galaxies comes from measurements of He I and H I recombination lines in H II regions, and the selection of results shown in Table 8.1 indicates that there are little if any He abundance differences among the nearby observed galaxies. Likewise, the few heavy-element abundance studies that have been made, particularly of H II regions in the Large and Small Magellanic Clouds, M 31, M 33, M 51, and M 101, indicate that these abundances do not differ greatly from the abundances in our own Galaxy. These studies, however, are not very precise, because weaker lines, such as [O III] $\lambda4363$, have not been measured, and therefore the temperature is not observationally determined. However, the similarity of the spectra shows that there are no gross composition differences between the H II regions observed to date in other galaxies and the H II regions in our Galaxy.

External galaxies furnish the opportunity to survey an entire galaxy, and an interesting study of M 33, M 51, and M 101 shows differences in the relative strengths of emission lines in H II regions, depending on the distance of the H II region from the center of the galaxy. In general, the [O III]/Hβ ratio increases outward in each of these galaxies, while the [N II]/Hα ratio decreases outward. Simple models seem to indicate that these differences

TABLE 8.1
Helium Abundance in Other Galaxies

Galaxy	H II Region	N_{He^+}/N_p	N_{He}/N_H
M 33	NGC 604	0.094	0.13
	NGC 604	0.134	0.17
M 101	NGC 5461	0.082	0.10
	NGC 5471	0.092	0.10
NGC 4449	Anon	0.078	0.10
LMC	NGC 2070	0.105	0.12
	NGC 2070	0.083	0.09
SMC	NGC 346	0.086	0.09

NOTE: Independent determinations in the same object are listed separately and give an idea of the range of uncertainty.

represent composition gradients in the galaxies, in the sense that the N/H abundance ratio decreases outward, while the O/H ratio decreases more slowly; and as a result T increases outward, though other possible interpretations also remain open.

Advances in detecting and accurately measuring weak spectral lines with photoelectric systems, such as image-tube scanners, will undoubtedly be applied to these problems, and we may hope to learn a good deal about possible composition variations in galaxies. A difficulty, however, is that we can study only the brightest H II regions in other galaxies, which are typically ionized by fairly large numbers of stars with a range of spectral types. The stars cannot be individually observed because of the distances of the galaxies, and the level of ionization in the nebula therefore cannot be computed from the properties of the star, but must be determined empirically from the observations themselves.

8.3   Distribution of H II Regions in Our Galaxy

The analogy with observed external galaxies, of course, strongly suggests that the H II regions in our own Galaxy are also concentrated to spiral arms. There is no doubt that H II regions are strongly concentrated to the galactic plane because, except for the very nearest, they are all close to the galactic equator in the sky. However, our location in the system and the strong concentration of interstellar dust to the galactic plane make it difficult to survey much of the Galaxy optically for H II regions and to determine their distances accurately. The surveys are carried out photographically in the light of Hα plus [N II] λλ6548, 6583, the most complete being the National Geographic Society-Palomar Observatory Sky Survey, made with the 48-inch Schmidt telescope, which (including its southern extension) covers all of the sky north of declination −33°. In this survey, the red plates were taken with a red Plexiglass filter and 103a-E plates, isolating a spectral region approximately 6000 A < λ < 6700 A, as described in Section 8.2. The Whiteoak extension to −45°, taken with a wider spectral range (5400 A < λ < 6700 A, defined by an amber Plexiglass filter) is also available. For the southernmost part of the sky, inaccessible from northern observatories, the available surveys were taken with smaller instruments, the best available at present being the Mount Stromlo Survey, made with an 8-inch F/1 Schmidt. However, the European Southern Observatory and the Science Research Council are carrying out and planning surveys of the entire southern sky with Schmidt telescopes comparable to the Palomar instrument, and these surveys will no doubt soon supersede the Mount Stromlo Survey.

The only accurate method of finding the distance of H II regions is the method of spectroscopic parallaxes, that is, spectral classification of the stars involved to find their absolute magnitudes and hence their distances. To determine the distance accurately requires highly accurate spectral classification, accurate color measurements to determine the interstellar extinction, and careful calibration of the relationships between spectral type, absolute magnitude, and intrinsic color.

Part of the problem in finding the distances of H II regions is simply identifying the exciting star or stars whose distance, when measured, will give the distance of the nebula. Although, for most of the nearby bright H II regions, the exciting O stars can easily be recognized from available spectral surveys, there are some, even among the nearest, in which the exciting star has not yet been identified with certainty. For instance, in NGC 7000, the North America Nebula, shown in Figure 8.2, the exciting star is probably the sixth magnitude O5 star HD 199579 near the middle of the photograph, but some workers have thought that this star may be simply a projected foreground object, and that other exciting stars may be hidden behind the dense obscuring cloud that runs across the H II region and divides North America from the Pelican. In more distant H II regions, the problem often is that the exciting star or stars cannot be identified among the many foreground and background stars projected on the nebula because there are no adequate spectral surveys of OB stars fainter than about twelfth magnitude. In fact, with an image-tube spectrograph plus a plus-counting photoelectric photometer, it is possible to determine the spectral types and colors of quite faint O stars once they have been found, and thus to measure

204

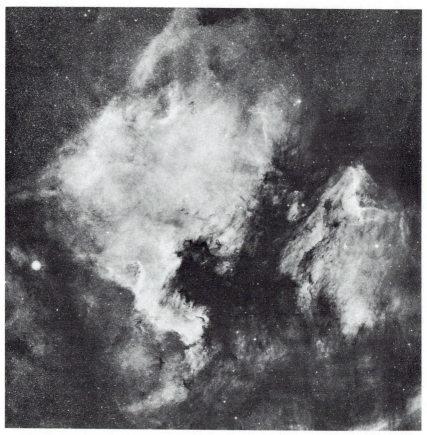

FIGURE 8.2
NGC 7000, North America Nebula (*to left*), and smaller IC 5067, Pelican Nebula (*to right*). Both are apparently parts of one large H II region, separated in the sky by foreground extinction of a dense interstellar cloud. HD 199579, the O5 star referred to in the text, is just over 1-1/4″ down from top edge, 2″ from right edge. Original plate, taken with the Palomar 48-inch Schmidt, with red filter, emphasizing Hα, [N II] λλ6548, 6583. (*Hale Observatories photograph.*)

their distances, but the main obstacle at present is the lack of spectral surveys in which to find the O stars to observe in detail.

The available distances of H II regions and OB star aggregates, plus a few very distant single stars, are plotted in Figure 8.3, which shows that sections of three spiral arms are well delineated from the optical measurements. Note, however, that the range of observation is small and that the distances of most of the H II regions plotted are less than 2 kpc from the sun, while the distance to the galactic center is approximately 10 kpc. Notice also that the extreme southern-hemisphere part of the galactic plane, which cannot be observed from the northern observatories and is therefore much less completely studied, appears on this map as a large blank sector between galactic longitudes approximately 210° and 330°.

A map of the southern-hemisphere results is shown in Figure 8.4. The figure shows that, around galactic longitude 290°, a spiral arm apparently runs off with the opposite inclination to the northern spiral arms. However, part of the problem is that there is a clear region in this direction, with very little interstellar extinction, so that it is possible to observe quite distant stars and nebulae in the longitude range 280° to 300°. If the distances of

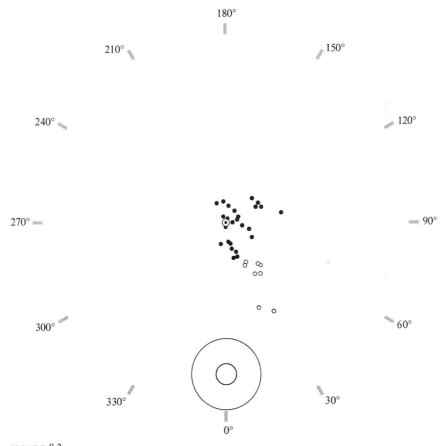

FIGURE 8.3
Northern-hemisphere H II regions plotted as solid circles in the plane of the Galaxy. The point ⊙ indicates the sun, and eight distant OB stars are plotted as open circles. Sections of three spiral arms can be seen: the outer (Perseus) arm, the local (Orion) arm, and the inner (Sagittarius) arm. Galactic longitudes are indicated, with 0° toward the galactic center, which is shown as two concentric circles at an adopted distance of 10 kpc from the sun.

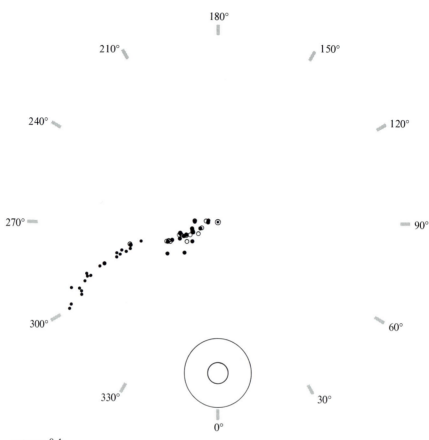

FIGURE 8.4
Southern-hemisphere H II regions in the plane of the Galaxy are plotted as solid circles; young star clusters are plotted as open circles; and distant stars are plotted as dots. The point ⊙ indicates the sun, and the galactic center is shown as two concentric circles at an adopted distance of 10 kpc from the sun.

these stars are not accurately known, relatively small percentage errors appear as rather large displacements, and this probably contributes to the apparent general spreading out of the spiral arm in this direction.

To date only the brightest O stars in the southern hemisphere have been studied, which means that in regions with higher interstellar extinction outside the galactic range 280° to 300°, only the nearer H II regions have been measured. It seems certain that, when spectral surveys for OB stars have been pushed to fainter magnitude limits, when the individual exciting stars of H II regions have been identified, and when their distances have been accurately determined, we shall have a much better picture of the spiral structure than we do at the present.

Optical studies of more distant H II regions in our Galaxy are rendered impossible by interstellar extinction. However, this is not a problem in the radio-frequency region, and the radio-continuum and recombination-line measurements have detected very distant H II regions. However, further study of these optically invisible nebulae is difficult because there is no direct way to measure their distances, since stars cannot be identified in them. The only method for finding the distance is then to use the observed radial velocity of the H II region, together with a model of the variation of galactic rotation velocity with distance. Such models, of course, have been widely used for the 21-cm H I observations. However, careful comparison of distances derived in this way with distances derived from spectroscopic classification of the identified stars in relatively nearby H II regions shows that the models are too simplified, and though the results they give are better than having no information whatsoever on the distance, they are not highly reliable.

Figure 8.5, a map of the entire Galaxy, shows H II regions observed by their recombination-line radiation. Notice that approximately 200 nebulae are included, which, of course, are the most luminous H II regions in our Galaxy and are more or less comparable with the 200 brightest H II regions that show on an Hα photograph of an external galaxy such as M 31. They are, on the average, significantly more luminous than the 30 or so H II regions, most of which are within 2 kpc of the sun, that are plotted in Figure 8.3. Note that the galactic rotation model does not uniquely determine the distance of H II regions closer to the center of the Galaxy than the sun, since the same measured radial velocity corresponds to two possible positions equally distant from the galactic center, one closer to the sun, on the sun's side of the perpendicular from the center of the Galaxy to the line from the sun through the nebula, and the other further from the sun, on the other side of this perpendicular. In Figure 8.5 each H II region has been plotted in both positions unless there is some additional information by which the ambiguity may be resolved. This ambiguity does not exist for H II regions more distant from the galactic center than the sun. However, for all H II regions, any errors in the kinematic model that links position in the Galaxy and velocity, and also any dispersion about this relationship, lead to errors in the derived distance. One other interesting aspect of these radio observations of distant H II regions is that they give the opportunity for measuring the He abundance in nebulae far from the sun (as indicated in Section 5.8), at the same time that their velocities are determined to estimate their distances.

It is also possible to get some information from the radio observations on the total amount of ionized gas in the Galaxy. This is most easily done from measurements of the radio-frequency continuum, which is the strongest radio-frequency emission from H II regions. Measurements must be made

208

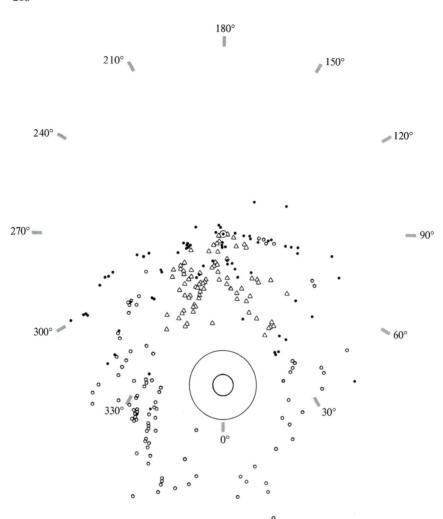

H II regions with distances determined from H 109α radial velocity and from a model of galactic rotation. H II regions closer to the galactic center than to the sun may be at either of two positions indicated by a triangle or a circle. If there is a clear reason to adopt one distance, the H II region is indicated by a dot and the other distance is not plotted at all.

at two well-separated frequencies, at both of which the Galaxy is optically thin, in order to separate the nonthermal synchrotron radiation (which increases toward lower frequency) from the thermal free-free radiation (which has a flux nearly independent of frequency). The free-free surface brightness in any direction gives the integral

$$T_{b\nu} = 8.24 \times 10^{-2} \, T^{-0.35} \nu^{-2.1} \int N_+ N_e \, ds \qquad (8.1)$$

(see equations 4.37 and 4.32). The mean effective length is defined by a galactic model, so the measurement gives the mean-square ion density averaged along the entire ray. The integral is, of course, proportional to the number of recombinations along the ray, and hence to the number of ionization processes, so it gives directly the number of ionizing photons absorbed in H II regions and in the lower-density gas between the nebulae. To derive the amount of ionized gas requires an estimate of its distribution along the ray, that is, of the clumpiness or filling factor. With an estimated density in the emitting regions $N_e \approx 5$ cm^{-3}, corresponding to a filling factor ranging from 0.1 to 0.01 depending on distance from the galactic center, such observations indicate a total mass of ionized gas in the galactic plane of order $4 \times 10^7 \, M_\odot$. This is only a small fraction of the total amount of gas determined from 21-cm H I observations, $3 \times 10^9 \, M_\odot$, which, in turn, is itself only a small fraction of the total mass of the Galaxy, $2 \times 10^{11} \, M_\odot$.

8.4   Stars in H II Regions

The existence of an H II region requires that there be one or more O stars within a region of reasonably high interstellar gas density, because only these stars emit enough photons with $h\nu > h\nu_0 = 13.6$ eV to ionize a sufficiently large amount of gas so that the nebula is an easily observable H II region. Often these O stars are in early type clusters containing relatively large numbers of B stars; for instance, the O stars that ionize NGC 6524 belong to the cluster NGC 6530, which contains at least 23 known B stars. It is known from theoretical stellar evolution calculations that O stars have a maximum lifetime of few $\times$ 10^6 yr, before they exhaust their central available H fuel and become supergiants; thus the clusters in which they occur must be recently formed star clusters. The luminosity function of such a cluster therefore provides an example of the luminosity function of a group of recently formed stars.

One type of low-luminosity star of which many examples have been found in nearby H II regions are T Tauri stars, which are main-sequence G and K stars that vary irregularly in light and have H and Ca II emission lines in their spectra. Only the nearest H II regions can be surveyed for these stars because of their intrinsically low luminosities, but many of them have been found in NGC 1976. On the other hand, T Tauri stars can also exist in the regions of high density of interstellar gas that is not ionized; for instance, many of these stars are also found in the Taurus dark nebulosity. Thus they are recently formed stars not necessarily directly connected with the formation of O stars.

An H II region first forms when an O star "turns on" in a region of high interstellar gas density. The star must have formed from interstellar matter, and observational evidence strongly suggests that a high density of interstellar matter is necessary before star formation can begin. Radio observations, in particular, have led to the discovery of many small, dense, "compact H II regions," with $N_e \sim 10^4$ cm^{-3}, in nebulae that are optically invisible because of high interstellar extinction.

Once a condensation has, probably as a result of turbulent motions, reached sufficiently high density to be self-gravitating, it contracts, heating up and radiating photons by drawing on the gravitational energy source. Once the star becomes hot enough for nuclear reactions to begin, it quickly stabilizes on the main sequence. It seems likely that many nebulae form as a result of density increases, perhaps in the collision of two or more lower-density interstellar clouds, and that in the resulting high-density condensation, star formation rapidly begins. For instance, observations show that NGC 1976 has a very steep density gradient, with the highest density quite near but not exactly coincident with the Trapezium, which includes the ionizing stars θ^1 Ori C and θ^1 Ori D.

After the O star or stars in a condensation "turn on," an R-type ionization front rapidly runs out into gas at a rate determined by the rate of emission of ultraviolet photons by the star(s). Ultimately, the velocity of the ionization front reaches the R-critical velocity, and at this stage the front becomes D-critical and a shock wave breaks off and runs ahead of it, compressing the gas. The nebula continues to expand and may develop a central local density minimum as a result of radiation pressure exerted on the dust particles in the nebula. Ultimately, the O star exhausts its nuclear energy sources and presumably becomes a supernova, though this is by no means certain. In any case, the nebula expanding away from the central star has drawn kinetic energy from the radiation field of the star, and this kinetic energy is ultimately shared with the surrounding interstellar gas. From the number of O stars known to exist, it is possible to think that a significant fraction of the interstellar turbulent energy is derived from the photoionization input of O stars communicated through H II regions, though there are many observational uncertainties in such a picture.

8.5   Molecules in H II Regions

Within the past few years, many interstellar molecular emission lines have been detected in the infrared and radio-frequency regions. The first interstellar molecule detected by its radio-frequency lines, OH, has been observed in many H II regions. The transition is between the two components of the ground $^2\Pi_{3/2}$ level that are split by Λ-type doubling; each component is further split by hyperfine interaction, so that there is a total of four lines with frequencies 1612, 1665, 1667, and 1720 MHz. A typical observed line in an H II region has a profile that may be divided into several components with different radial velocities, and the relative strengths of these components often vary in times as short as a few months. Many of the individual components have narrow line profiles, nearly complete circular polarization or strong linear polarization, and high brightness temperature (in some cases, $T_{b\nu} \geqslant 10^9 {}^\circ K$). All of these characteristics are strong evidence for maser action resulting from nonthermal populations of the individual molecular levels. Further, many of the OH sources are also observed by their H_2O radio-frequency emission lines, though not all OH sources have these lines. Some of the OH radiation comes from extended regions in H II regions, but a large fraction of it and all of the H_2O radiation comes from very small, bright sources within the H II regions. Studies made with very long base-line interferometers show that the OH emission usually occurs in clusters of small sources, the clusters having sizes typically of $1''$, while the individual sources within the cluster have diameters of order $0''.005$ to $0''.5$.

In the H II regions that have been studied to date, the OH sources tend to occur in areas of strong interstellar extinction. Since OH molecules would be rapidly dissociated in the strong ultraviolet radiation field within an H II region, it seems likely that the sources are small, dense condensations that are optically thick to ionizing radiation, so that their interiors are shielded by the surface layers of gas and dust. In these regions of high dust density, molecules can be formed by collisions of atoms and simpler molecules with dust particles, surface reactions on the dust particles, and subsequent escape. The molecules are excited by collisions with other molecules and atoms, and presumably also by resonance fluorescence due to ultraviolet radiation with $h\nu < h\nu_0 = 13.6$ eV, which penetrates into the clouds.

In addition to OH and H_2O, the molecules CO, CN, CS, HCN, H_2CO, and CH_3OH have all been detected in NGC 1976, which is the best studied H II region, and several of them have been detected in other H II regions as well. Some of these molecules are undoubtedly concentrated in or are escaping from small, dense condensations in the nebula itself, while others, particularly CO, are spread along the line of sight between the sun and the nebula. The observational study of molecules is quite new, and little is known about their distribution outside the galactic center and H II regions on which the observers have tended to concentrate because they were the sources in which the first molecules were discovered.

It can confidently be expected that many new discoveries will be made from radio-frequency and microwave measurements of molecular lines. In H II regions, it appears likely that these lines will be particularly useful in probing dense, cool condensations, but in addition there may perhaps be other completely new types of structures that they will reveal.

References

The distribution of H II regions "like beads on a string" along the spiral arms of M 31, and the photographic survey by which he discovered this distribution, are clearly described by

Baade, W. 1951. *Pub. Obs. Univ. Michigan* **10**, 7.

The detailed results of this survey, including coordinates of the H II regions and maps of their distribution projected on the sky and in the deduced plane of M 31 itself, are given in

Baade, W., and Arp, H. 1964. *Ap. J.* **139**, 1027.

Arp, H. 1964. *Ap. J.* **139**, 1045.

A photographic survey of a large number of other galaxies for H II regions is described and published in detail, respectively, in

Hodge, P. W. 1967. *A. J.* **72**, 129.

Hodge, P. W. 1969. *Ap. J. Supp.* **18**, 73.

The general features of the distribution of H II regions in spiral galaxies and irregular galaxies, respectively, deduced from this survey, are given in

Hodge, P. W. 1969. *Ap. J.* **155**, 417.

Hodge, P. W. 1969. *Ap. J.* **156**, 847.

Early spectroscopic measurements of radial velocities of H II regions in M 31 are given by

Babcock, H. W. 1939. *Lick Obs. Bull.* **19**, 41.

Mayall, N. U. 1951. *Pub. Obs. Univ. Michigan* **10**, 19.

The most recent and complete study is:

Rubin, V. C., and Ford, W. K. 1970. *Ap. J.* **159**, 379.

An early paper on the spectra of H II regions in several nearby galaxies is:

Aller, L. H. 1942. *Ap. J.* **95**, 52.

A recent very complete paper is:

Searle, L. 1971. *Ap. J.* **168**, 327.

This paper discusses abundance variations with position of H II regions in M 33, M 51, and M 101. Abundance determinations from H II regions in other galaxies are scattered through the literature, but a summary containing many references to the original work is:

Osterbrock, D. E. 1970. *Q. J. R. A. S.* **11**, 199.

(Table 8.1 is taken from this reference.) The most complete abundance study of an H II region in an external galaxy (NGC 604 in M 33) is

Aller, L. H., Czyzak, S. J., and Walker, M. F. 1968. *Ap. J.* **151**, 491.

The National Geographic Society-Palomar Observatory Sky Survey is available in the form of photographic prints of the individual fields. Brief descriptions of it have been published in

Minkowski, R. L., and Abell, G. O. 1963. *Basic Astronomical Data,* ed. K. A. Strand. Chicago: University of Chicago Press, p. 481.

Lund, J. M., and Dixon, R. S. 1973. *P. A. S. P.* **85**, 230.

The Mount Stromlo Survey was issued as a separate publication:

Rodgers, A. W., Campbell, C. T., Whiteoak, J. B., Bailey, H. H., and Hunt, V. O. 1960. *An Atlas of* Hα *Emission in the Southern Milky Way.* Canberra: Mount Stromlo Observatory.

Earlier southern surveys for H II regions include:

Gum, C. S. 1955. *Mem. R. A. S.* **67**, 155.

Bok, B. J., Bester, M. J., and Wade, C. M. 1955. *Proc. Am. Acad. Art. Sci.* **86**, 9.

Wide-angle photographs showing the brightest H II regions close to the galactic equator are reproduced in

Osterbrock, D. E., and Sharpless, S. 1952. *Ap. J.* **115**, 89.

Mapping the spiral arms by optical determinations of the distances of OB stars in H II regions is discussed in

Morgan, W. W., Sharpless, S., and Osterbrock, D. E. 1952. *A. J.* **57**, 3.

Morgan, W. W., Whitford, A. E., and Code, A. D. 1953. *Ap. J.* **118**, 318.

(Figure 8.3 is adapted from this reference.)

Morgan, W. W., Code, A. D., and Whitford, A. E. 1965. *Ap. J. Supp.* **2**, 41.

Bok, B. J., Hine, A. A., and Miller, E. W. 1970. *Spiral Structure of Our Galaxy* (IAU Symposium No. 38), ed. W. Becker and G. Contoupoulos. Dordrecht: D. Reidel, p. 44.

(Figure 8.4 is adapted from this reference.)

Bok, B. J. 1971. *Highlights of Astronomy,* ed. C. de Jager. Dordrecht: D. Reidel, p. 63.

Bok, B. J. 1972. *American Scientist* **60**, 709.

The problem of finding very faint OB stars from objective prism or other spectroscopic surveys is very well described by

Morgan, W. W. 1951. *Pub. Obs. Univ. Michigan* **10**, 33.

Morgan, W. W., Meinel, A. B., and Johnson, H. M. 1954. *Ap. J.* **120**, 506.

Schulte, D. H. 1956. *Ap. J.* **123**, 250.

Comparisons of optically determined distances of nebulae, from spectroscopic classification of their involved stars, with determinations based on measured velocities and a standard model of galactic rotational velocities, have been made by

Miller, J. S. 1968. *Ap. J.* **151**, 473.

Georgelin, Y. P., and Georgelin, Y. M. 1971. *Ast. and Ap.* **12**, 482.

The most complete radio-recombination-line surveys of H II regions in our Galaxy at the present time are H 109α (λ = 6 cm) surveys, reported in

Reifenstein, E. C., Wilson, T. L., Burke, B. F., Mezger, P. G., and Altenhoff, W. J. 1970. *Ast. and Ap.* **4**, 357.

Wilson, T. L., Mezger, P. G., Gardner, F. F., and Milne, D. K. 1970. *Ast. and Ap.* **6**, 364.

(Figure 8.5 is adapted from the latter reference, but includes the results of both these surveys.) The continuum survey from which the entire amount of ionized gas in the Galaxy is derived is

Westerhout, G. 1958. *Bull. Ast. Inst. Netherlands* **14**, 261.

Compact H II regions are described in

Mezger, P. G., Altenhoff, W., Schraml, J., Burke, B. F., Reifenstein, E. C., and Wilson, T. L. 1967. *Ap. J.* **150,** L157.

Mezger, P. G. 1968. *Interstellar Ionized Hydrogen,* ed. Y. Terzian. New York: Benjamin, p. 33.

The discovery of the OH molecule in interstellar space by its radio-frequency absorption lines is reported in

Weinreb, S., Barrett, A. H., Meeks, M. L., and Henry, J. C. 1963. *Nature* **200,** 829.

Some excellent later reviews, which contain many detailed references to work on molecules in H II regions, are:

Robinson, B. J., and McGee, R. X. 1967. *Ann. Rev. Astr. and Astrophys.* **5,** 183.

Rank, D. M., Townes, C. H., and Welch, W. J. 1971. *Science* **174,** 1083.

Star formation is a very large subject in itself. A very good book, consisting of reviews and papers by many experts, is:

O'Connell, D. J. K., ed. 1958. *Stellar Populations,* Amsterdam: North Holland Publishing Co.

Some excellent more recent summaries are:

Spitzer, L. 1968. *Nebulae and Interstellar Matter,* ed. B. M. Middlehurst and L. H. Aller. Chicago: University of Chicago Press, p. 1.

McNally, D. 1971. *Reports Progress Phys.* **34,** 71.

Strom, S. E. 1972. *P. A. S. P.* **84,** 745.

Herbig, G. H. 1970. *Spectroscopic Astrophysics,* ed. G. H. Herbig. Berkeley: University of California Press, p. 237.

Good symposia on interstellar matter and young stars and on NGC 1976, the Orion Nebula, including the stars in it, are published in

Pecker, J. C., ed. 1966. *Transactions of the IAU.* **12B,** pp. 412, 443. London: Academic Press.

de Jager, C., ed. 1971. *Highlights of Astronomy* **2,** 335. Dordrecht: D. Reidel. The last reference is a symposium on interstellar molecules.

9

Planetary Nebulae

9.1 Introduction

The previous chapters have summarized the ideas and methods of nebular research, first treating nebulae from a static point of view, then adding the effects of motions and of dust particles. In the first seven chapters, many references have been made to actual planetary nebulae, but only a fraction of the known results have been discussed. This final chapter will complete the discussion of what has been learned about planetary nebulae. First their space distribution in the Galaxy and their galactic kinematics are summarized; then what is known about the evolution of the nebulae and of their central stars, including ideas on the origin of planetary nebulae, is discussed. This leads naturally to a discussion of the rate of mass return of interstellar gas to the Galaxy from planetary nebulae and their significance in galactic evolution. Finally, the chapter summarizes what is known about planetary nebulae in other galaxies.

9.2   Space Distribution and Kinematics of Planetary Nebulae

Except for the brightest classical planetary nebulae, which were identified by their finite angular sizes, planetary nebulae are discovered photographically by objective-prism surveys or by direct photography in a narrow spectral region around a strong emission line or lines, such as [O III] or Hα and [N II] $\lambda\lambda6583, 6548$. An objective-prism survey tends to discover small, bright, high-surface-brightness objects, while direct photography tends to discover nebulae with large angular sizes, even though they have low surface brightness. These surveys, of course, penetrate only the nearer regions of the Galaxy, because of the interstellar extinction by dust concentrated to the galactic plane. A total of about 700 planetary nebulae are known, and their angular distribution on the sky, as shown in Figure 9.1, exhibits fairly strong concentration to the galactic plane, but not so strong as H II regions, and strong concentration to the center of the Galaxy. It must be remembered that in this map the concentration to the galactic equator and to the galactic center would undoubtedly be more extreme if it were not for the interstellar extinction, which preferentially suppresses the more distant planetaries.

The only directly measured trigonometric parallax of a planetary nebula that is reasonably well determined is $0.''042 \pm 0.''011$ for NGC 7293. The estimate of the probable error, which depends on the internal consistency of the individual measurements at one observatory, is a lower limit to the actual uncertainty, as can be seen from comparison of parallaxes of other planetary nebulae measured at two or more observatories. One planetary nebula, NGC 246, has a late-type main-sequence companion star, and the spectroscopic parallax of this star implies a distance of between 360 and 480 pc. In a few nebulae, comparison of the tangential proper motion of expansion with the measured radial expansion velocity gives distance estimates, but these are very uncertain because, as discussed in Chapter 6, the velocity of expansion varies with position in the nebula, and in some nebulae the apparent motion of the outer boundary may be the motion of an ionization front rather than the mass motion, which is measured by the Doppler effect. Measurements of proper motions of about 35 planetary-nebula central stars are available, which give a statistical mean parallax for the group. Finally, some planetaries are known in the Magellanic Clouds, whose distances are known independently. This is all the direct information that exists on the distances of planetary nebulae, and it is not sufficient for a study of the space distribution of these objects, so it is necessary to use other less direct and correspondingly less accurate methods of distance estimation.

The basic assumption of the indirect method, commonly known as the Shklovsky distance method, is that all planetary-nebula shells are completely

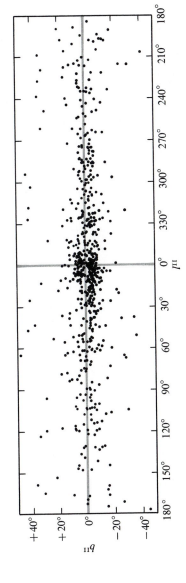

FIGURE 9.1
Distribution of planetary nebulae in the sky, plotted in galactic coordinates. Note concentration to galactic equator and to galactic center.

ionized and have approximately the same mass, so that as they expand, their mean electron densities decrease and their radii r_N increase according to the law

$$\frac{4\pi}{3} r_N{}^3 N_e = \text{const.} \tag{9.1}$$

Hence measurement of the electron density of any planetary nebula makes it possible to determine its radius, and if its angular radius is then measured, its distance directly follows. The electron density cannot be directly measured except in planetaries with [O II] or [S II] lines, but it is possible instead to measure the mean Hβ surface brightness (or intensity) of the nebula $I_{\mathrm{H}\beta}$:

$$I_{\mathrm{H}\beta} \propto N_e N_p r_N \propto N_e{}^2 r_N \propto N_e{}^{5/3} \propto r_N{}^{-5}. \tag{9.2}$$

There is some check of this method in that the expected relation between Hβ surface brightness and N_e is verified for those planetary nebulae in which the [O II] lines have been measured. Solving for the distance D of the nebula,

$$D = \frac{r_N}{\phi} \propto \frac{(I_{\mathrm{H}\beta})^{-1/5}}{\phi}$$

$$\propto (\pi F_{\mathrm{H}\beta})^{-1/5} \phi^{-3/5} \tag{9.3}$$

where ϕ is the angular radius of the nebula, and $\pi F_{\mathrm{H}\beta}$ is its measured flux in Hβ (corrected for interstellar extinction) at the earth, so that $I_{\mathrm{H}\beta} \propto \pi F_{\mathrm{H}\beta} \phi^{-2}$.

For instance, if we adopt a spherical planetary-nebula model in which a fraction ϵ, the "filling factor" of the volume, contains gas with uniform density N_p, N_e its mass is

$$M_N = \frac{4\pi}{3} N_p r_N{}^3 (1 + 4y) m_{\mathrm{H}}, \tag{9.4}$$

where y is the abundance ratio $N_{\mathrm{He}}/N_{\mathrm{H}}$. If we write the electron density $N_e = N_p (1 + xy)$, so that x gives the fractional ionization of He$^+$ to He^{++}, the expression for the flux from the nebula

$$\pi F_{\mathrm{H}\beta} = \frac{\frac{4\pi}{3} N_p N_e r_N{}^3 \epsilon \, \alpha_{\mathrm{H}\beta}^{\mathrm{eff}} h\nu_{\mathrm{H}\beta}}{4\pi D^2} \tag{9.5}$$

can be expressed in terms of $r_N = \phi D$ and solved for

$$D = \left[\frac{3}{16\pi^2} \frac{M_N{}^2}{m_{\mathrm{H}^2}} \frac{(1 + xy)}{(1 + 4y)^2} \epsilon \, \alpha_{\mathrm{H}\beta}^{\mathrm{eff}} h\nu_{\mathrm{H}\beta} \right]^{1/5}$$

$$\times (\pi F_{\mathrm{H}\beta})^{-1/5} \phi^{-3/5}. \tag{9.6}$$

This equation is then used to determine the distance of a planetary nebula from measured values of its flux and angular size, the first factor in square brackets being treated as a constant that is determined from the nebulae of known distance discussed previously. Alternatively, any other recombination line (such as Hα) or the radio-frequency continuum could be used— the equations are always similar to equation (9.6) and only the numerical value of the constant is different.

Note that in equation (9.6), the distance depends only weakly on the assumed mass of ionized gas. At the time of writing, the best available calibration, which depends largely on the planetary nebulae in the Magellanic Clouds, corresponds to the value $M_N = 0.2\ M_\odot$, and this can be taken as the average mass of ionized gas in an average planetary nebula. It is, of course, only rather poorly determined, for the very reason that the distance depends only weakly upon it. A catalogue of planetary nebulae using this distance calibration is available, and all distances quoted in the present chapter come from it.

One check on the method is that, since planetary nebulae are expected to expand with constant velocity, the number within each range of radius between r_N and $r_N + dr_N$ should be proportional to dr_N, and this test is in fact approximately fulfilled by the planetaries within a standard volume near the sun, corrected for incompleteness, in the range of radii $0.1\ \text{pc} \leqslant r_N \leqslant 0.7\ \text{pc}$. The larger nebulae with lower density and correspondingly lower surface brightness are more difficult to discover, and objects with $r_N > 0.7\ \text{pc}$ are essentially undetectable. On the other hand, below a definite lower limit $r_N \leqslant r_1$ set by the ultraviolet luminosity of the central star, the nebula is so dense that it is not completely ionized, and consequently its true ionized mass is smaller than assumed under the constant-mass hypothesis. Thus its true distance is smaller than that calculated using a constant coefficient in equation (9.6), and we therefore should expect an apparent underabundance of nebulae with calculated radius $r_N \leqslant r_1$, and a corresponding excess of nebulae with $r_N \geqslant r_1$. This effect does occur in the statistics of observed planetary nebula sizes, with $r_1 \approx 0.07\ \text{pc}$, which gives us some confidence in the indirect photometric distance method.

However, the derived distances of the planetary nebulae can only be approximate and correct statistically, because direct photographs show that their forms, internal structures, and ionization all have considerable ranges. The central assumption of a uniform-density spherical model is too idealized to represent real nebulae accurately. The figures of many planetary nebulae are more nearly axisymmetric toroidal than spherical, as, for example, NGC 6720, shown in Figure 9.2. In addition, the statistical arguments can never eliminate the possibility that a small fraction of the planetary nebulae (say, 10 or 20 percent) are objects with completely different natures than the other planetaries, but with a similar appearance in the sky. Nevertheless, the indirect Shklovsky distance method is the only method we have for

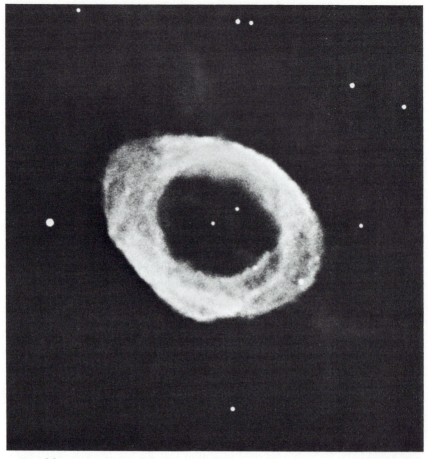

FIGURE 9.2
NGC 6720, the Ring Nebula in Lyra, a classical planetary nebula. Original photograph was taken with 120-inch reflector, red filter, and 103a-E plate, emphasizing Hα, [N II] $\lambda\lambda$6548, 6583. (*Lick Observatory photograph.*)

measuring the distances of planetaries and drawing statistical conclusions about their space distribution and evolution.

The radial velocities of many planetary nebulae have also been measured and exhibit a relatively high-velocity dispersion. The measured radial velocities are plotted against galactic longitude in Figure 9.3. It can be seen that the radial velocities of the planetary nebulae in the direction $l \approx 90°$ tend to be negative, and in the direction $l \approx 270°$ they tend to be positive, which shows that the planetary nebulae are "high-velocity" objects, that is, they belong to a system that actually has a considerably smaller rotational velocity about the galactic center than the sun, so they appear to us to be moving, on the average, in the direction opposite the sun's galactic rotation. Figure 9.3 also shows the high dispersion of velocities in the direction of the galactic center. On the basis of the relative motion of the system of planetaries with respect to the local circular velocity, planetary nebulae are generally classified as old Population I objects, but not as such outstanding high-velocity objects as Extreme Population II.

Though, because of interstellar extinction, it is not possible to survey the entire Galaxy for planetary nebulae, the discovery statistics should be fairly complete up to a distance in the galactic plane of about 1000 pc, and the observed total number of planetaries within a cylinder of radius 1000 pc centered on the sun, perpendicular to the galactic plane, is 41, corresponding to a surface density of planetary nebulae (projected on the galactic plane) near the sun of $1.3 \times 10^{-5} \, pc^{-2}$. The statistics are increasingly incomplete at large distances because of interstellar extinction, though some planetaries are known at very great distances at high galactic latitudes. It is only possible to find the total number of planetaries in the whole Galaxy by fitting the local density to the model of their galactic distribution based on stars of approximately the same kinematical properties. The number of planetary nebulae in the whole Galaxy found in this way is approximately 4×10^5.

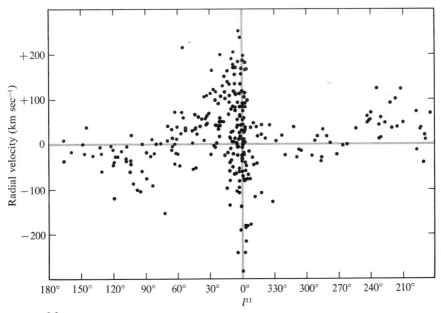

FIGURE 9.3

Observed radial velocities of planetary nebulae, reduced to local standard of rest, plotted against galactic longitude.

This number is, of course, not nearly so well determined as the more nearly directly observed local surface density.

The height distribution of the planetary nebulae in the Galaxy may be derived from the known distances of planetaries. The average distance from the galactic plane $\langle|z|\rangle$ of the planetary nebulae within 1000 pc projected distance from the sun in the galactic plane is about 150 pc, while the root-mean-square distance $(\langle z^2 \rangle)^{1/2} \approx 215$ pc. This is a fairly strong concentration to the galactic plane, approximately the same as the Intermediate Population I of Oort. It should be noted, however, that there is a tremendous range in properties of the planetaries, for in spite of their rather strong concentration to the galactic plane, one nebula, Haro's object HA4-1, is at a height $z \approx 1.1 \times 10^4$ pc from the plane.

9.3   The Origin of Planetary Nebulae and the Evolution of Their Central Stars

Observations of the distances of planetary nebulae give information on the properties of their central stars, for the distance of a nebula, together with the measured apparent magnitude of its central star, gives the absolute magnitude of the star. Furthermore, the effective temperature of the central star, or at least a lower limit to this temperature, can be found by the Zanstra method described in Section 5.7. The idea is that the measured flux in a recombination line such as Hβ is proportional to the number of recombinations in a nebula and hence to the whole number of ionizing photons absorbed in the nebula; if the nebula is optically thick to ionizing radiation, this number is, in turn, equal to the whole number of ionizing photons emitted by the central star. Comparison of the number of ionizing photons with the number of photons in an optically observed wavelength band gives T_*, the effective temperature of the central star, which in turn determines the bolometric correction, that is, the difference between the visual and bolometric magnitudes of the star. If the nebula is optically thin so that all the ionizing photons are not absorbed, this method gives only a lower limit to the relative number of ionizing photons emitted by the central star and hence a lower limit to T_*.

If He II lines are observed in a nebula, the same method may be used for the He$^+$-ionizing photons with $h\nu \geqslant 4h\nu_0 = 54$ eV. Almost all nebulae with He II are optically thick to He$^+$-ionizing radiation; nebulae in which [O I] lines are observed are almost certainly optically thick to H^0-ionizing radiation. In this way, a luminosity-effective temperature diagram of the central stars of planetary nebulae can be constructed; the results are as shown in Figure 9.4. The individual uncertainties are great, as can be seen from the error bars, because of the difficulties of making photometric measurements of these faint stars immersed in nebulosity, and also because

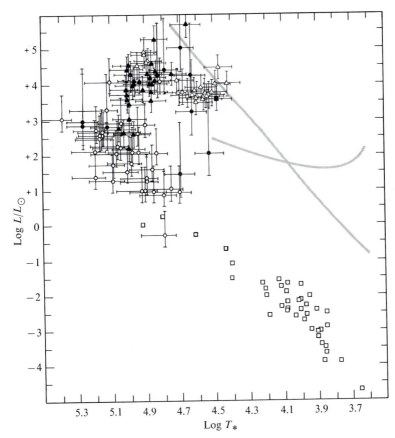

FIGURE 9.4

Observed luminosity-effective temperature diagram for central stars of planetary
nebulae (*circles*) and white dwarfs (*squares*). Bars give the estimated uncertainty
for each planetary-nebula star. Gray lines represent the positions of the
Population I main sequence and the Population II horizontal branch.

of uncertainties as to the completeness of absorption of the ionizing photons.
Nevertheless, it is clear that the effective temperatures range up to
$2 \times 10^{5}{}^{\circ}$K; they are as high as, if not higher than, temperatures determined
for any other types of stars. The luminosities are much higher than the sun,
and at their brightest they are as luminous as many supergiant stars.

Furthermore, it is possible to attach a time since the planetary was "born,"
that is, since the central star lost the shell that became the planetary nebula,
to each point plotted on this diagram, since the radius of each planetary
is known from its distance, and the radius, together with the mean expansion
velocity, that measures the time since expansion began. When this is done
it is seen that the youngest planetaries are those with central stars around

$L \approx 3 \times 10^2 \, L_\odot$, $T \approx 50,000°$K, while the somewhat older planetaries have central stars around the maximum $L \approx 2 \times 10^4 \, L_\odot$ of Figure 9.4; and still older planetaries have successively less luminous central stars. This must mean that the central stars of planetary nebulae evolve around the path from O to D shown in Figure 9.5, a schematic L-T_* diagram, in the same time that the nebula expands from essentially zero radius at formation to a density so low that it disappears at $r_N \approx 0.7$ pc. This time, with a mean expansion velocity of 20 km sec^{-1}, is about 3.5×10^4 yr, much shorter than almost all other stellar-evolution times, and shows that the planetary-nebula phase is a relatively short-lived stage in the evolution of a star.

At D the observed L and T_* correspond to a stellar radius $R \approx 0.025 \, R_\odot$, so in the final stage of a planetary nebula, the nebular shell expands and merges with the interstellar gas, while the central star becomes a white dwarf. Separation between the shell and remnant star must occur at very nearly a composition discontinuity, because the nebula has approximately normal abundance of H, while a white-dwarf star can have almost no H at all except at its very surface. If such a star had a normal abundance of H, it would be producing energy by nuclear reaction at a much higher rate and would not be a stable white dwarf. As Figures 9.4 and 9.5 show, the oldest planetary-nebula central stars are in the white-dwarf region of the L-T_* diagram,

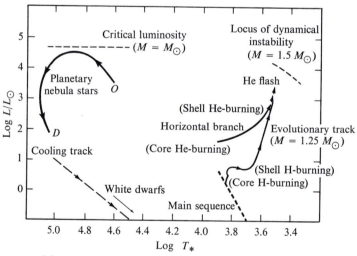

FIGURE 9.5

Schematic luminosity-effective temperature diagram showing the evolution of planetary-nebula central stars from birth, at O to the white-dwarf stage, near the end of their observed evolutionary track. The diagram also shows earlier evolutionary stages, including the zero-age main sequence, and calculated evolutionary tracks up to the red-giant branch and the horizontal branch, as well as later evolution down through the white-dwarf region.

just above the region in which many white dwarfs lie. In this region a white dwarf's radius, which is fixed by its mass, remains constant, and the star simply radiates its internal thermal energy, becoming less luminous along a cooling line

$$L = 4\pi R^2 \sigma T_*^4. \tag{9.7}$$

Two of the most interesting problems in the study of planetary nebulae are the nature of their progenitors and the process by which the nebula is formed. The velocity of expansion of the nebular shell, approximately 20 km sec^{-1}, is so low in comparison with the velocity of escape of the present planetary-nebula central stars, of order 1000 km sec^{-1}, that it is quite unlikely that any impulsive process that throws off an outer shell from the star would provide just a little more energy than the necessary energy of escape. On the other hand, the velocities of expansion are comparable with the velocities of escape from extreme red giants, which suggests that the shell is ejected in the red-giant or supergiant stage.

Let us recall the evolution of a fairly low-mass star with $M_\odot < M < 4M_\odot$ and its track in the Hertzsprung-Russell diagram as shown schematically in Figure 9.5. After contraction to the main sequence, hydrogen burning continues in the central core for most of the star's lifetime, until all the H in the core is exhausted. The nearly pure He core then begins to contract, H burning begins in a shell source just outside the core, and the star evolves into the red-giant region and develops a deep outer convection zone that increases in depth as the H-burning shell moves outward in mass and as the core becomes denser and hotter. When the central temperature $T_c \approx 1 \times 10^8 °$K is reached near the red-giant tip, He burning begins in the helium flash, and the star rapidly moves to lower L and higher T_* in the Hertzsprung-Russell diagram, to a position on the "horizontal branch." It appears that the planetary-nebula shell is not ejected at the time of the helium flash because both theory and observation show that the stars immediately after their first excursion to the red-giant tip are horizontal-branch stars, not incipient white dwarfs. A horizontal-branch star initially burns He in its central core and H in its outer shell source, and the direction of its evolution immediately after arriving on the horizontal branch (toward larger or smaller T_*) depends on the relative strengths of these two energy sources. After the star burns out all the He at its center, it consists of a central C + O-rich core, an intermediate He-rich zone, and an outer H-rich region. There are He-burning and H-burning shell sources at the inner edges of the two latter regions, and the star evolves with increasing L and decreasing T_* toward the red-giant tip again. This evolution would terminate when a central temperature $T_c \approx 6 \times 10^8 °$K is reached and C burning begins, but apparently before this happens the ejection of the planetary nebula occurs.

The most likely mechanism by which the shell is ejected from the super-giant star is dynamical instability against pulsations. When this instability occurs in very extended red giants, the energy stored in ionization of H and He may be large enough so that the total energy of the outer envelope of the star is positive, and the pulsation amplitude can then increase without limit, lifting the entire envelope off the star and permitting it to escape completely. Calculations of this process show that the boundaries between stable and dynamically unstable stars in the mass range $M_\odot \leqslant M \leqslant 3M_\odot$ all occur around $L \approx 10^4 L_\odot$ for cool red giants near the region in the L-T_* diagram occupied by observed long-period variables. Detailed calculations have been made showing that stars in this region have envelopes with positive total energies corresponding to velocities at infinity (if the material were expanded adiabatically) of about $30 \, \mathrm{km \ sec^{-1}}$. Furthermore, these calculations show that if the outer part of the envelope is removed, then the remaining model with the same luminosity and the same mass in its burned-out core, but with a smaller mass remaining in the envelope, is more unstable, so that if mass ejection starts by this process, it probably will continue until the entire envelope, down to the bottom of the H-rich zone, is ejected. The process is stopped at this level by the discontinuity in density due to the discontinuity in composition. Thus, according to these ideas, the mass left in the planetary-nebula central star is the mass in the burned-out core of the parent star when it first becomes unstable. The available calcula-tions show that a star of $0.8 \, M_\odot$ produces a remnant star of $0.6 \, M_\odot$ and a nebular shell of $0.2 \, M_\odot$; a $1.5 \, M_\odot$ star produces a remnant star of $0.8 \, M_\odot$ and a shell of $0.7 \, M_\odot$; and a $3 \, M_\odot$ star produces a remnant star of $1.2 \, M_\odot$ and a shell of $1.8 \, M_\odot$, while stars with $M \gtrsim 4M_\odot$ begin burning C explosively and presumably become supernova before they become dynamically un-stable. Stars with $M \lesssim 0.6 \, M_\odot$ cannot become dynamically unstable at all, but of course, in any event, stars with $M \lesssim 0.9 \, M_\odot$ or so have not yet evolved to the planetary-nebula stage in the lifetime of the Galaxy. The presence of dust in planetary nebulae suggests that the material in the shell came from a cool stellar atmosphere and somewhat strengthens the evidence for the evolution of planetary nebulae from red-giant stars.

Another mechanism that has been suggested for the expulsion of a plane-tary-nebula shell is radiation pressure in a small, hot, blue star not too different from the young planetary-nebula central stars themselves. In this process the acceleration is not instantaneous, since the force is a long-range force, and the final small velocity at infinity can be understood. To see what is involved, we can split the ordinary hydrostatic equilibrium equation

$$\frac{dP}{dr} = \frac{dP_{\mathrm{gas}}}{dr} + \frac{dP_{\mathrm{rad}}}{dr} = -\frac{GM_r \rho}{r^2} \tag{9.8}$$

into gas and radiation pressure terms, and if we substitute

$$\frac{dP_{\text{rad}}}{dr} = -\frac{\kappa\rho L_r}{4\pi r^2 c},$$

(9.9)

this becomes

$$\frac{dP_{\text{gas}}}{dr} = -\frac{GM_r\rho}{r^2}\left(1 - \frac{\kappa L_r}{4\pi cGM_r}\right).$$

(9.10)

The gas pressure gradient becomes positive, corresponding to a repulsive force, if the second term within the parentheses is larger than unity. This occurs in the outer part of a star with $M = M_\odot$ and electron-scattering opacity at a critical luminosity $L \approx 4 \times 10^4 L_\odot$, which is actually not far above the observed luminosities of the brightest planetary nebula central stars, as can be seen from Figure 9.5. Note from equation (9.10) that the critical luminosity is proportional to the mass of the star.

This radiation-pressure mechanism has been investigated in detail, and the result is that in models in which L increases rapidly, in times of order 10^3 yr or less, shells can be driven off to become planetary nebulae. In this process the entire H-rich envelope would be ejected from the nebula, leaving behind the inner burnt-out core as the remnant star because of the discontinuity in opacity. The main difficulty with accepting this mechanism for the formation of planetary nebulae is that no commonly accepted evolutionary tracks lead naturally to the assumed progenitors.

Models of the rapidly evolving central star after the planetary nebula has been ejected have been calculated by several astrophysicists. The essential features of the evolution can be reproduced qualitatively by a star with a degenerate C + O core, or a degenerate core consisting of an inner C + O region and a small outer He region, together with possibly a very small-mass, H-rich envelope. The inner core hardly evolves at all as the envelope burns H rapidly in a shell source and contracts, while the He region, if it is small enough, contracts without igniting. Neutrino processes rapidly cool the interior, and the entire star becomes a white dwarf and then simply cools at constant radius. Some problems remain with the time scale and the details of the track, but the general features of the observed track are approximately reproduced.

9.4 Mass Return from Planetary Nebulae

The gas in a planetary-nebula shell, of course, ultimately mixes with and thus returns to interstellar matter. The present rate of this mass return, calculated from the local density of planetaries derived from distance measurements in the way explained in Section 9.2, is 7.8×10^{-10} planetary shells $pc^{-2} yr^{-1}$ projected on the galactic plane near the sun, or $1.5 \times 10^{-10} M_\odot pc^{-2} yr^{-1}$. This rate is larger than the rate of mass return observed from any other type of object except for the estimated rate $6 \times 10^{-10} M_\odot pc^{-2} yr^{-1}$ from long-period variables, which is derived from infrared observations, together with very specific assumptions about the nature and optical properties of the dust particles in these stars. The rate of mass loss by planetaries is, however, considerably smaller than the estimated total rate of mass loss from all stars to interstellar matter, based on a model of the Galaxy, the luminosity function or distribution of star masses, and calculations of their lifetimes, which gives approximately $10 \times 10^{-10} M_\odot pc^{-2} yr^{-1}$.

Integrated over the entire Galaxy, the rate of planetary nebula "deaths" is approximately 25 per year; in other words, the rate of mass return to interstellar space is approximately $5 M_\odot yr^{-1}$. This may be compared with the rate of mass return to interstellar matter from supernovae, using the rate of one supernova per 25 yr estimated for our Galaxy. Approximately one-third of the galactic supernovae are Type 1 and two-thirds are Type 2; the mass in either type of shell is not at all well known, but is crudely estimated from tentative theoretical ideas as $5 M_\odot$ for Type 2 supernovae and much less for Type 1. According to these estimates, the rate of mass return to interstellar space from supernovae would be of order $0.07 M_\odot yr^{-1}$ integrated over the whole Galaxy, less by a factor of approximately 100 than the rate of mass return from planetaries. Presumably, however, the material in supernova shells has been processed much more thoroughly by nuclear reactions and has abundances that are considerably more extreme than those within a typical planetary-nebula shell.

The death rate of planetary nebulae may also be compared with the birth rate of white-dwarf stars derived from observational data on the local density of white dwarfs in each color range. This quantity is only very poorly known because of the intrinsic faintness of the white dwarfs and the small region that may therefore be surveyed, but the best estimate is that, near the sun, new white dwarfs form at a rate of about $2 \times 10^{-12} pc^{-3} yr^{-1}$. This estimate is very close to the birth rate of planetary nebulae, which is the same as their death rate, $2.5 \times 10^{-12} pc^{-3} yr^{-1}$, near the galactic plane.

Thus the overall conclusion is that the very uncertain observational data do not disagree with the idea that a large fraction of the stars with original masses in the range 1 to $4 M_\odot$ end their evolution by throwing off shells and becoming white dwarfs. The question then naturally arises whether all stars in the stated mass range go through this evolution. It seems unlikely that they do, because pre- and postnovae are hot blue stars in the same general region of the Hertzsprung-Russell diagram as planetary-nebula central stars, and apparently have masses of the general order of $1 M_\odot$, but do not occur anywhere in the evolutionary track leading to planetary nebulae outlined previously. Observational evidence strongly suggests that all novae are relatively close binaries, so a possible interpretation is that all single stars (or members of wide binary pairs) evolve through the planetary-nebula stage, while all members of close binaries evolve through the nova stage, in which numerous small outbursts, rather than a single planetary episode, are responsible for dispersion of the outer part of the star. Obviously, much further research, both observational and theoretical, is needed on this problem.

9.5 Planetary Nebulae in Other Galaxies

Planetary nebulae are not as luminous as giant H II regions, and they are correspondingly less easily observed in other galaxies. However, the Large and Small Magellanic Clouds are close enough so that numerous planetaries have been discovered in them by objective-prism surveys, and individual slit spectra have been obtained for many of these planetaries. The distance of the Clouds, about 6×10^4 pc, is large enough so that most of the planetary nebulae in our Galaxy would not be resolved at this distance, and indeed none of the planetaries in the Clouds have been resolved in direct photographs. However, some of the very largest galactic planetary nebulae, for instance, an object like NGC 7293 with diameter about 0.5 pc, should just barely be resolved with apparent angular diameters of 1″ to 2″ if they were in the Magellanic Clouds. Some planetary nebulae probably will be resolved in the future under conditions of fine seeing with large southern-hemisphere telescopes. This is important because the calibration of the distance scale for planetary nebulae discussed in Section 9.2 depends heavily on the planetaries in the Magellanic Clouds. Observed radial velocities of the planetaries in the Magellanic Clouds also help to define their population characteristics.

In more distant galaxies, four planetary nebulae were identified in M 31 many years ago by Baade, who compared plates taken in [O III] λλ4959, 5007 with plates taken in the nearby continuum. These nebulae must be among the very brightest planetary nebulae in M 31, but they still are so faint that very little information is available on them, except for spectrograms showing a few lines. However, the availability of narrow-band interference filters and image tubes, which make photography possible to much fainter light levels, may make possible the discovery of many more planetary nebulae in M 31 and other nearby galaxies.

230

References

General references on planetary nebulae include:

Seaton, M. J. 1960. *Reports Progress Phys.* **23**, 313.

Osterbrock, D. E. 1964. *Ann. Rev. Astr. and Astrophys.* **2**, 95.

Minkowski, R. 1965. *Galactic Structure,* ed. A. Blaauw and M. Schmidt. Chicago: University of Chicago Press, p. 321.

(Figure 9.3 is based on this reference.)

Osterbrock, D. E., and O'Dell, C. R. 1968. *Planetary Nebulae* (IAU Symposium No. 34). Dordrecht: D. Reidel.

Perek, L., and Kohoutek, L. 1967. *Catalogue of Galactic Planetary Nebulae.* Prague: Czechoslovak Academy of Sciences.

This is an excellent source that contains practically all observational data on planetary nebulae published through mid-1966.

General references on surface distribution and kinematics, respectively, include:

Minkowski, R., and Abell, G. O. 1963. *P. A. S. P.* **75**, 488.

(Figure 9.1 is based on this reference.)

Minkowski, R. 1965. *Galactic Structure.* ed. A. Blaauw and M. Schmidt. Chicago: University of Chicago Press, p. 321.

(Figure 9.3 is based on this reference.)

Some references on the distance scale include the following:

Shklovsky, I. S. 1956. *Russian Astron. J.* **33**, 515.

Osterbrock, D. E. 1960. *Ap. J.* **131**, 541.

O'Dell, C. R. 1962. *Ap. J.* **135**, 371.

Seaton, M. J. 1966. *M. N. R. A. S.* **132**, 113.

Seaton, M. J. 1968. *Astrophys. Letters* **2**, 55.

Cahn, J. H., and Kaler, J. B. 1971. *Ap. J. Supp.* **22**, 319.

The first reference describes the idea of obtaining distances of planetaries from the hypothesis that they all contain the same amount of ionized gas, and the second describes the observational check of the correlation between electron density and surface brightness that is predicted on the basis of this hypothesis. The last reference is the most complete existing catalogue of distances of planetary nebulae, based on the indirect method, and the numerical values quoted in the text are based on it. The second and third of these references also discuss the derived temperatures of the central stars, as well as the evolution of central stars of planetary nebulae, as do the following three references:

Osterbrock, D. E. 1966. *Stellar Evolution,* ed. A. G. W. Cameron and W. Stein. New York: Plenum Press, p. 381.

O'Dell, C. R. 1968. *Planetary Nebulae* (IAU Symposium No. 34). Dordrecht: D. Reidel, p. 361.

(Figure 9.4 is based on this reference.)

Salpeter, E. E. 1971. *Ann. Rev. Astr. and Astrophys.* **9**, 127.

The basic idea that the low velocities of expansion of planetary nebulae indicate that their progenitors are probably extended red giants was first stated by Shklovsky in the reference mentioned previously in this section and was further developed by

Abell, G. O., and Goldreich, P. 1966. *P. A. S. P.* **78**, 232.

The idea that red giants may evolve to planetaries was earlier proposed from astrophysical arguments by

Menzel, D. H. 1946. *Physica* **12**, 768.

The idea that pulsational instability is the mechanism, and that escape of the shell occurs because the outer part of an extended red giant has positive total energy, was suggested by

Lucy, L. B. 1967. *A. J.* **72**, 813.

Roxburgh, I. W. 1967. *Nature* **215**, 838.

This mechanism was investigated and shown to occur in detailed calculations by

Paczynski, B., and Ziolkowski, J. 1968. *Acta Astronomica* **18**, 255.

Scott, E. H. 1973. *Ap. J.* **180**, 487.

Properties of white dwarfs are summarized in

Weidemann, V. 1968. *Ann. Rev. Astr. and Astrophys.* **6**, 351.

Models and evolutionary tracks of planetary-nebula central stars have been calculated by

Rose, W. K., and Smith, R. L. 1970. *Ap. J.* **159**, 903.

Vila, S. C. 1970. *Ap. J.* **162**, 605.

Paczynski, B. 1971. *Acta Astronomica* **21**, 417.

The subject is very well summarized by

Salpeter, E. E. 1971. *Ann. Rev. Astr. and Astrophys.* **9**, 127.

The radiation-pressure mechanism of shell ejection has been worked out by

Faulkner, D. J. 1970. *Ap. J.* **162**, 513.

A related continuous-flow mechanism was worked out by

Finzi, A., and Wolf, R. 1971. *Astron. and Ap.* **16**, 418.

Summarizing references on stellar evolution are numerous, but two excellent ones are:

Iben, I. 1967. *Ann. Rev. Astr. and Astrophys.* **5**, 571.

Iben, I. 1971. *P. A. S. P.* **83**, 697.

The return of matter from planetary nebulae to interstellar matter, formation of white dwarfs, and so on, are discussed by

O'Dell, C. R. 1968. *Planetary Nebulae* (IAU Symposium No. 34). Dordrecht: D. Reidel, p. 361.

Torres-Peimbert, S., and Peimbert, M. 1971. *Bol. Obs. Tonantzintla Tacubaya* **6**, 101.

Osterbrock, D. E. 1973. *Mem. Société Roy. Sci. Liège* **5**, 391.

Observational work on the planetary nebulae in the Magellanic Clouds is given by

Webster, B. L. 1969. *M. N. R. A. S.* **143**, 79.

Webster, B. L. 1969. *M. N. R. A. S.* **143**, 97.

Webster, B. L. 1969. *M. N. R. A. S.* **143**, 113.

Glossary of Physical Symbols

The symbols used in this book are listed here with their physical significance and the section (indicated without parentheses) or equation (indicated by parentheses) in which they first appear. Dummy mathematical variables and symbols used only once are not listed. Note that in some cases the same symbol has been used for two widely differing quantities, usually one from quantum mechanics and the other from stellar or nebular astronomy.

ROMAN

a	Radius of a dust particle	7.3
$A_{i,j}$	Radiative transition probability between upper level i and lower level j	2.2
a_0	Bohr radius $h^2/4\pi^2 me^2$	(2.4)
a_T	Threshold absorption cross section for an arbitrary atom or ion	(2.31)
A_λ	Albedo of a dust particle at wavelength λ	(7.9)
a_ν	Absorption cross section per atom	(2.1)
b_j	Deviation from thermodynamic equilibrium factor	(4.6)
$B_\nu(T)$	Planck function at frequency ν	(4.34),(4.36)
C	Constant giving amount of interstellar extinction along a ray for use with natural logarithms	(7.3)
c	Constant giving amount of interstellar extinction along a ray for use with logarithms to base 10	(7.6)

c	Velocity of light	(2.14)
$C(i,j)$	Collisional transition rate per ion in level i from level i to level j, with temperature exponential factored out	(5.3)
$C_{i,j}$	Probability population of upper level i is followed by population of lower level j	4.2,(4.10)
c_0	Isothermal velocity of sound in $H°$ region	6.3
c_1	Adiabatic velocity of sound in H^+ region	6.3
D/Dt	Time derivative following an element of volume	(6.1)
E	Emission measure	(4.32)
E	Internal kinetic or thermal energy per unit mass	(6.5)
e	Electron charge (absolute value)	2.2
E_p	Proton-emission measure	(5.14),(5.15)
f	Fraction of excitations to a level leading to emission of a particular line photon	5.9
f_{ij}	f-value of a line between a lower level i and an upper level j	(4.45)
$f(v)$	Maxwell-Boltzmann distribution function	(2.6)
F_ν	Flux of radiation divided by π; $\pi F_\nu =$ flux of radiation	(2.2)
$F_{\nu d}$	Flux of diffuse radiation divided by π; $\pi F_{\nu d} =$ diffuse flux	(2.10)
$F_{\nu s}$	Flux of stellar radiation divided by π; $\pi F_{\nu s} =$ stellar flux	(2.11)
G	Energy input rate due to photoionization	(3.1),(3.8)
g_{ff}	Gaunt factor for free-free emission	(4.22)
g_ν	Frequency dependence of $H°$ 2-photon emission coefficient	(4.29)
h	Planck's constant	(2.1)
I_ν	Specific intensity of radiation	(2.9)
$I_{\nu d}$	Specific intensity of diffuse radiation field	(2.10)
$I_{\nu s}$	Specific intensity of stellar radiation field	(2.10)
j_{ij}	Emission coefficient in a line resulting from a radiative transition from an upper level i to a lower level j	(4.12)
J_ν	Mean specific intensity of radiation; $J_\nu = 1/4\pi \int I_\nu \, d\omega$	(2.1)
j_ν	Emission coefficient in continuum	(2.9)
k	Boltzmann constant	(2.6)

k_{ol}	Line absorption coefficient at center of a line	(4.43),(4.44)
k_{vL}	Line absorption coefficient at frequency v corrected for stimulated emission	(4.42)
k_{vl}	Line-absorption coefficient at frequency v	(4.42)
L	Luminosity of star; $L = \int L_v dv$	7.6
L	Orbital angular momentum quantum number	2.2
L_c	Energy loss rate due to collisionally excited radiation	(3.23),(3.30)
L_{FF}	Energy loss rate due to free-free emission	(3.14)
Lj	Lyman line from upper level j to lower level l	(4.13)
L_R	Energy loss rate due to recombination	(3.3),(3.8)
L_v	Luminosity of star per unit frequency interval	(2.2)
M	Mach number	(6.19)
m	Electron mass	2.2
m_D	Mass of a dust particle	7.3
m_H	Mass of proton or hydrogen atom	(4.13)
M_V	Absolute visual magnitude	1.5
$M_\odot$	Solar mass	1.4
n	Principal quantum number	2.2
$N_c(i)$	Critical electron density for collisional de-excitation of level i	(2.23),(3.31)
n_{cL}	Principal quantum number above which collisions dominate distribution of atoms among levels with different angular momentum L	4.2
N_D	Number density of dust particles per unit volume	(7.9)
N_e	Electron density	1.3
N_H	Hydrogen density; $N_H = N_{H^\circ} + N_p$	2.1
N_{H°	Neutral hydrogen atom density	(2.1)
N_{He}	Helium density; $N_{He} = N_{He^\circ} + N_{He^+} + N_{He^{++}}$	(2.28)
N_{He°	Neutral helium atom density	(2.21)
N_{He^+}	Singly ionized helium density	2.4
N_j	Density of atoms in level j	(4.1)
N_p	Proton density	(2.1)
p	Gas pressure	(6.1)
$P_{i,j}$	Probability that population of upper level i is followed by a directive radiative transition to lower level j	(4.8)

$q_{i,j}$	Collisional transition rate from level i to level j per particle in level i per colliding particle per unit volume per unit of time	(2.22)
$Q(X)$	Number of ionizing photons for element X emitted by star; for example:	

$$Q(\mathrm{H}^\circ) = \int_{\nu_0}^{\infty} \frac{L_\nu}{h\nu}\, d\nu \qquad (2.19)$$

Q_λ	Extinction cross section of a dust particle at wavelength λ	(7.8)
R	Stellar radius	(2.2)
r	Distance from star	(2.2)
r	Ratio of line to continuum brightness temperature	(5.17)
r_1	Strömgren radius or critical radius of H^+ zone	(2.19)
r_2	Radius of He^+ zone corresponding to Strömgren radius of H^+ zone	(2.27)
r_3	Radius of He^{++} zone corresponding to Strömgren radius of H^+ zone	(2.29)
S_ν	Source function in equation of transfer	(5.19)
T	Absolute thermodynamic temperature	1.3
$T_{b\nu}$	Brightness temperature at frequency ν	(4.37)
T_C	Brightness temperature in the radio-frequency continuum	5.6
T_c	Color temperature of continuum radiation	7.3
T_D	Temperature of a dust particle	(7.5)
T_i	Initial temperature of newly created photoelectrons	(3.2)
T_L	Brightness temperature at the center of a radio-frequency line	5.6
T_*	Stellar effective temperature; $L = 4\pi R^2 \sigma T_*^4$	1.3
U	Internal kinetic or thermal energy per unit volume	(6.4)
v	Electron velocity	(2.5)
$\mathbf{v}$	Velocity vector	(6.1)
v_D	D-critical velocity	(6.24)
v_R	R-critical velocity	(6.23)
X_n	Ionization potential of level n	(4.4),(4.5)
Z	Charge on a nucleus or dust particle in units of electron charge	7.3
Z	Nuclear charge in units of charge of proton	(2.4)

GREEK

α	Recombination coefficient	(2.1)
α_A	Recombination coefficient summed over all levels	(2.7)
α_B	Recombination coefficient summed over all levels above ground level; $\alpha_B = \alpha_A - \alpha_1$	(2.18)
α_1	Recombination coefficient to level i	(2.5)
$\alpha_{ij}^{\mathrm{eff}}$	Effective recombination coefficient for emission of a line resulting from a radiative transition from upper level i to lower level j	(4.14)
$\alpha(X, T)$	Recombination coefficient to the species X at temperature T	2.1
β	Correction to line-absorption coefficient due to maser effect	(5.18)
$\beta_A(X, T)$	Effective recombination coefficient for recombination energy loss of species X at temperature T	(3.4)
β_i	Effective recombination coefficient to level i for recombination energy loss	(3.4)
γ_ν	Frequency dependence of continuum-emission coefficients	(4.23),(4.24)
$\epsilon(x)$	Escape probability of a photon emitted at a point x in a nebula	4.5
κ_C	Absorption coefficient per unit volume in continuum	5.6
κ_L	Line absorption coefficient per unit volume at center of a line	(5.11)
λ	Wavelength	1.1
λ_{ij}	Wavelength corresponding to transition from upper level i to lower level j	4.2
λ_0	Threshold wavelength for ionization of $H°$; $\lambda_0 = c/\nu_0$	2.2
μ	Mean atomic or molecular weight for particle	(6.2)
ν	Frequency; $\nu = c/\lambda$	(2.1)
ν_{ij}	Frequency corresponding to a transition from upper level i to lower level j	3.5
ν_T	Threshold frequency for ionization of an arbitrary atom or ion	(2.31)
ν_0	Threshold frequency for ionization of $H°$; $h\nu_0 = 13.6\ eV$	2.1
ν_1	Threshold frequency for ionization of H-like ion of charge Z; $\nu_1 = Z^2\nu_0$	(2.4)

ν_2	Threshold frequency for ionization of He$^\circ$; $h\nu_2 = 24.6$ eV	2.4
ξ	Fraction of neutral H; $\xi = N_{H^\circ}/(N_{H^\circ} + N_p)$	2.1
ρ	Gas density	(6.1)
$\sigma_{i,j}(v)$	Collisional cross section for transition from level i to level j for relative collision velocity v	(2.22)
$\sigma_i(X, v)$	Recombination cross section to level i of species X electrons with velocity v	(2.5)
τ_C	Optical depth in the continuum	(5.10)
τ_{cL}	Optical depth at the center of a line corrected for stimulated emission	(5.10)
τ_{cl}	Optical depth at the center of a line	4.5
τ_i	Mean lifetime of excited level i	(2.3)
τ_L	Contribution to the optical depth at the center of a line due to the line alone	(5.10)
τ_0	Optical depth at frequency ν_0	2.3
$\tau_\lambda(i)$	Optical depth at wavelength λ along ray to a star or nebula i	(7.1)
τ_ν	Optical depth at frequency ν	(2.12)
ϕ_i	Flux of ionizing photons	(6.13),(6.14)
χ	Threshold energy for an excitation process	(2.22)
ω_i	Statistical weight of level i	(3.15)
$\Omega(i, j)$	Collision strength between levels i and j	(3.15)

MISCELLANEOUS

*	Used to denote a quantity under conditions of thermodynamic equilibrium	5.6
$\partial/\partial t$	Partial time derivative at a fixed point in space	(6.1)

Milne Relation between Capture and Photoionization Cross Sections

The Milne relation expresses the capture cross section to a particular level (with threshold v_T) in terms of the absorption cross section from that level, and is based on the principle of detailed balancing or microscopic reversibility. According to this principle, in thermodynamic equilibrium each microscopic process is balanced by its inverse. Thus, in particular, recombination (spontaneous plus induced) of electrons with velocity in the range between v and $v + dv$ is balanced by photoionization by photons with frequencies in the range between v and $v + dv$, where

$$\frac{1}{2} mv^2 + hv_T = hv,$$

so

$$mv \, dv = h \, dv.$$

The rate of induced downward radiative transitions (induced recombinations in this case) in thermodynamic equilibrium is always just $e^{-hv/kT}$ times the rate of induced upward transitions (photoionizations in this case), so the equilibrium equation may be written

spontaneous recombination rate $= (1 - e^{-hv/kT})$ photoionization rate

or, in the notation of Chapter 2,

$$N_e N(X^{+i+1})\sigma(v) f(v) \, dv = (1 - e^{-hv/kT})N(X^{+i})\frac{4\pi B_v(T)}{hv} a_v \, dv.$$

Substituting the thermodynamic equilibrium relations for the Maxwell-

Boltzmann distribution function,

$$f(v) = \frac{4}{\sqrt{\pi}} \left(\frac{m}{2kT}\right)^{3/2} v^2 e^{-mv^2/2kT},$$

the Planck function,

$$B_\nu(T) = \frac{2h\nu^3}{c^2} \frac{1}{e^{h\nu/kT} - 1},$$

and the Saha equation,

$$\frac{N(X^{+i+1})N_e}{N(X^{+i})} = \frac{2\omega_{i+1}}{\omega_i} \left(\frac{2\pi mkT}{h^2}\right)^{3/2} e^{-h\nu_T/kT},$$

we obtain the Milne relation,

$$\sigma(v) = \frac{\omega_i}{\omega_{i+1}} \frac{h^2\nu^2}{m^2c^2v^2} a_\nu.$$

Though the preceding equation was derived using arguments from thermodynamic equilibrium, it is a relation between the recombination cross section at a specific v and the absorption cross section at the corresponding ν. It can also be derived quantum mechanically and depends only on the fact that the matrix element between two states is independent of their order.

In particular, if the photoionization cross section a_ν can be represented by the interpolation formula

$$a_\nu = a_T \left[\beta \left(\frac{\nu}{\nu_T}\right)^{-s} + (1 - \beta) \left(\frac{\nu}{\nu_T}\right)^{-(s+1)} \right],$$

it follows that the recombination cross section can be written

$$\sigma(v) = \frac{\omega_i}{\omega_{i+1}} \frac{h^2\nu^2}{m^2c^2v^2} a_T \left[\beta \left(\frac{\nu}{\nu_T}\right)^{-s} + (1 - \beta) \left(\frac{\nu}{\nu_T}\right)^{-(s+1)} \right].$$

Substituting, the recombination coefficient to the level X^{+i},

$$\alpha(X^{+i}, T) = \int_0^\infty v\sigma(v) f(v) \, dv,$$

becomes

$$\alpha(X^{+i}, T) = \frac{4}{\sqrt{\pi}} \frac{\omega_i}{\omega_{i+1}} \left(\frac{m}{2kT}\right)^{3/2} e^{h\nu_T/kT} \frac{h^3\nu^3}{m^3c^2} a_T$$

$$\times \left[\beta E_{s-2}\left(\frac{h\nu_T}{kT}\right) + (1 - \beta)E_{s-3}\left(\frac{h\nu_T}{kT}\right) \right],$$

if s is an integer, where E_n is the exponential integral function. If s is nonintegral, $E_n(x)$ must be replaced in this formula by $x^{n-1}\Gamma(1 - n, x)$.

Escape Probability of a Photon Emitted in a Spherical Homogeneous Nebula

Let $\tau = \kappa R$ be the optical radius of the nebula (see Figure A2.1), where R is the linear radius. Consider a ray making an angle θ to the outward normal; the total optical length along this ray is $\tau_\theta = 2\tau \cos \theta$, so if ϵ is the (constant)-emission coefficient per unit volume per unit solid angle per unit time, the emergent intensity in this direction is

$$I(\theta) = \int_0^{\tau_\theta} \epsilon \, e^{-t} \, ds = \frac{\epsilon}{\kappa} (1 - e^{-2\tau \cos \theta}),$$

where s is the linear coordinate along the ray, and $t = \kappa s$ is the optical-length coordinate. Therefore, the outward flux per unit area per unit time is

$$\pi F = 2\pi \int_0^{\pi/2} I(\theta) \cos \theta \sin \theta \, d\theta$$

$$= \frac{\pi \epsilon}{2\kappa \tau^2} [2\tau^2 - 1 + (2\tau + 1)e^{-2\tau}].$$

To find the escape probability, this expression may be compared with the flux the nebula would emit in the absence of any absorption, which is simply given by the total emission in the volume divided by the area:

$$\pi F(0 \text{ absorption}) = \frac{4\pi \epsilon \, \dfrac{4\pi}{3} R^3}{4\pi R^2} = \frac{4\pi \epsilon R}{3}.$$

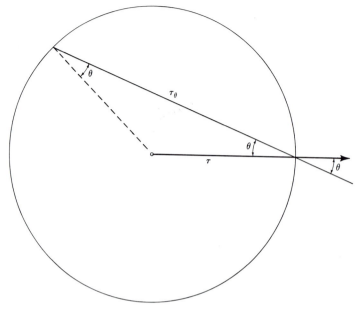

Cross section through the center of a spherical homogeneous nebula of radius r, optical radius $\tau = \kappa R$, showing a ray of optical length τ_θ making an angle θ with respect to the outward normal.

Thus the escape probability is

$$p(\tau) = \frac{\pi F}{\pi F(0 \text{ absorption})}$$

$$= \frac{3}{4\tau}\left[1 - \frac{1}{2\tau^2} + \left(\frac{1}{\tau} + \frac{1}{2\tau^2}\right)e^{-2\tau}\right].$$

It may be verified that as $\tau \to 0, p(\tau) \to 1$, while as $\tau \to \infty, p(\tau) \to 3/4\tau$, which is the contribution due to the escape of photons from the surface layer only.

Names and Numbers of Nebulae

Though most nebulae are ordinarily referred to by their numbers in the New General Catalogue (NGC) or the Index Catalogues, a few of the brighter ones are sometimes referred to in published papers by their Messier numbers. NGC numbers are used throughout the present book. The H II regions and planetary nebulae with Messier numbers (M) are listed in Table A.1, together with their NGC numbers and their common names, which are also sometimes used in published papers.

TABLE A.1
Identification of Messier Numbers

	M	NGC	Common name(s)
H II Regions			
	8	6523	Lagoon
	16	6611	—
	17	6618	Omega, Horseshoe
	20	6514	Trifid
	42	1976	Orion
	43	1982	Part of Orion complex
Planetary nebulae			
	27	6853	Dumbbell
	57	6720	Ring
	76	650–651	—
	97	3587	Owl
Supernova remnant			
	1	1952	Crab
Galaxies			
	31	224	Andromeda
	32	221	Round companion to Andromeda
	33	598	Triangulum
	51	5194	Whirlpool
	81	3031	—
	101	5457	—

Emission Lines of
Neutral Atoms

Collisionally excited emission lines of several neutral atoms are observed in gaseous nebulae, particularly [O I] $\lambda\lambda6300$, 6364, [N I] $\lambda\lambda5198$, 5200, [Mg I] $\lambda\lambda4562$, and Mg I] 4571. These emission lines arise largely from the transition regions or ionization fronts at the boundaries between H^+ and H^0 regions. As was described in Section 5.9, the relative strengths of the neutral-atom lines therefore provide information on the transition regions and on the dense neutral condensations in nebulae. The emission coefficients for these lines may be calculated by the methods described in Section 3.5, using the transition probabilities and collision strengths in the following tables. Note, however, that the collision strengths of neutral atoms are zero at the threshold and vary rapidly with energy; their mean values, defined by equation (3.20), must therefore be used. These mean values have been calculated for the most important collision strengths and are listed in Tables A.4 and A.5.

TABLE A.2
Transition Probabilities for [O I]

Transition	Transition probability (sec^{-1})	Wavelength (A)
$^1D_2-^1S_0$	1.34	5577.4
$^3P_2-^1S_0$	3.7×10^{-4}	2958.4
$^3P_1-^1S_0$	6.7×10^{-2}	2972.3
$^3P_2-^1D_2$	5.1×10^{-3}	6300.3
$^3P_1-^1D_2$	1.6×10^{-3}	6363.8
$^3P_0-^1D_2$	1.1×10^{-6}	6391.5
$^3P_1-^3P_0$	1.7×10^{-5}	—
$^3P_2-^3P_0$	1.0×10^{-10}	—
$^3P_2-^3P_1$	9.0×10^{-5}	—

TABLE A.3
Transition Probabilities for [N I] and Mg I]

Atom	Transition	Transition probability (sec^{-1})	Wavelength (A)
[N I]	$^4S_{3/2}-^2D_{3/2}$	1.6×10^{-5}	5197.9
[N I]	$^4S_{3/2}-^2D_{5/2}$	6.9×10^{-6}	5200.4
[Mg I]	$^1S_0-^3P_2^0$	2.8×10^{-4}	4562.5
Mg I]	$^1S_0-^3P_1^0$	4.3×10^2	4571.1

TABLE A.4
Collision Strengths for O^0

T	$\Omega(^3P,^1D)$	$\Omega(^3P,^1S)$	$\Omega(^3P_2,^3P_1)$	$\Omega(^3P_2,^3P_0)$	$\Omega(^3P_1,^3P_0)$
6,000	0.24	0.024	0.089	0.022	0.033
8,000	0.32	0.032	0.120	0.030	0.046
10,000	0.39	0.039	0.149	0.037	0.058
12,000	0.46	0.045	0.178	0.045	0.069
15,000	0.56	0.055	0.220	0.058	0.085
20,000	0.71	0.069	0.286	0.082	0.112

TABLE A.5
Collision Strengths for N^0 and Mg^0

T	N^0		Mg^0
	$\Omega(^4S,^2D)$	$\Omega(^2D_{3/2},^2D_{5/2})$	$\Omega(^1S,^3P)$
6,000	0.37	0.19	1.2
8,000	0.47	0.24	1.4
10,000	0.55	0.29	1.6
12,000	0.63	0.34	1.7
15,000	0.73	0.40	1.8
20,000	0.87	0.49	2.0

References

The transition probabilities for [O I] and [N I] in Tables A.2 and A.3 are taken from the compilation in

Garstang, R. H. 1968. *Planetary Nebulae* (I.A.U. Symposium No. 34), ed. D. E. Osterbrock and C. R. O'Dell. Dordrecht: D. Reidel, p. 143.

The transition probabilities for Mg I] and [Mg I] (the present text follows the convention of using a single bracket for electric-dipole intercombination lines, and both brackets for electric quadrupole and magnetic dipole and quadrupole lines) in Table A.3 are taken from

Wiese, W. L., Smith, M. W., and Miles, B. M. 1969. *Atomic Transition Probabilities* **2,** *Sodium through Calcium.* Washington, D.C.: Government Printing Office, p. 25.

These references list references to the original work, much of it done by Garstang himself.

The collision strengths used to form the mean values in Tables A.4 and A.5 were calculated by

Saraph, H. E. 1973. *J. Phys. B.* (Atom. Molec. Phys.) **6,** L243 (O^0).
Fabrikant, I. I. 1974. *J. Phys. B.* (Atom. Molec. Phys.) **7,** 91 (Mg^0).
Berrington, K., Burke, P. G., and Robb, W. D. *J. Phys. B.* (Atom. Molec. Phys.). To be published (N^0).

I am indebted to Dr. Saraph for sending me her results for O^0 at additional energies beyond those given in the reference, to Dr. Robb for sending me the unpublished results for N^0, and to all the authors for permitting me to include them in this book.

Index